T0143168

Higher and Colder

Higher and Colder

A History of Extreme Physiology and Exploration

VANESSA HEGGIE

The University of Chicago Press
Chicago and London

The University of Chicago Press, Chicago 60637

The University of Chicago Press, Ltd., London

© 2019 by The University of Chicago

All rights reserved. No part of this book may be used or reproduced in any manner whatsoever without written permission, except in the case of brief quotations in critical articles and reviews. For more information, contact the University of Chicago Press, 1427 E. 60th St., Chicago, IL 60637.

Published 2019

Printed in the United States of America

28 27 26 25 24 23 22 21 20 19   1 2 3 4 5

ISBN-13: 978-0-226-65088-3 (cloth)

ISBN-13: 978-0-226-65091-3 (e-book)

DOI: https://doi.org/10.7208/chicago/9780226650913.001.0001

Library of Congress Cataloging-in-Publication Data

Names: Heggie, Vanessa, author.

Title: Higher and colder: a history of extreme physiology and exploration / Vanessa Heggie.

Description: Chicago; London: The University of Chicago Press, 2019. | Includes bibliographical references and index.

Identifiers: LCCN 2018061119 | ISBN 9780226650883 (cloth: alk. paper) | ISBN 9780226650913 (e-book)

Subjects: LCSH: Cold regions—Physiological effect. | Extreme environments—Physiological effect.

Classification: LCC GB642 .H44 2019 | DDC 612/.014465—dc23

LC record available at https://lccn.loc.gov/2018061119

♾ This paper meets the requirements of ANSI/NISO Z39.48-1992 (Permanence of Paper).

To my parents, Debe & Alex, for everything

CONTENTS

ABBREVIATIONS

AMEE — American Mount Everest Expedition (1963)

AMREE — American Medical Research Expedition to Everest (1981)

AMS — acute mountain sickness

BAS — British Antarctic Survey

DIPAS — (India) Defense Institute of Physiology and Allied Sciences

ECG — electrocardiogram

EPO — erythropoietin

FIDS — Falkland Islands Dependencies Survey (later BAS)

IBEA — International Biomedical Expedition to the Antarctic (1980–81)

IBP — International Biological Program (1964–74)

ICAO — International Civil Aviation Organization

IGY — International Geophysical Year (1957–58)

IMP — integrating motor pneumotachograph

INPHEXAN — International Physiological Expedition to Antarctica (1957)

IOC — International Olympic Committee

MRC — Medical Research Council

NIMR — (UK) National Institute of Medical Research

OEI, II, III — Operation Everest

ONR — (US) Office for Naval Research

P_{O_2}, P_{CO_2} — partial pressure of oxygen, carbon dioxide

RAF — (UK) Royal Air Force

RGS — (UK) Royal Geographical Society

SCAR — Scientific Committee on Antarctic Research

TAE — Commonwealth Trans-Antarctic Expedition (1955–58)

Introduction: Higher and Colder

During the twentieth century, humans went in unprecedented numbers to the hottest, coldest, and highest points on the earth's surface. Many went to fulfill colonial, imperial, or military needs, but others went for pleasure, or for personal or national glory. None of these motivations were exclusive, and many were combined with a desire to "do science" in extreme environments. Scientists accompanied, assisted, organized, and led these expeditions. Geophysicists, geographers, and astronomers were particularly well represented in these spaces, but there were also biologists and ecologists counting penguins, collecting butterflies, and hunting yetis, and there were physiologists and biomedical scientists studying the explorers themselves. This book is a history of those biomedical scientists and physiologists whose interest in extreme environments was predicated not on the environment itself, or its organic or inorganic features, but on embodied human encounters with extremes of temperature, altitude, and living conditions. These researchers used exploration as an opportunity to study the human body at its physical limits, and in turn provided the guidance and the technology that allowed bodies to climb higher and to trek farther across icy or sandy deserts.

This book is not a chronological list of expeditionary scientific work—those studies already exist[1]—but rather a thematic exploration of the work of extreme physiologists in the twentieth century. It will prove two contradictory facts about this sort of biomedical work: first, that it is an extraordinary form of scientific practice, undertaken in unusual environments with unique challenges; and second, that despite its uniqueness, it functions as an excellent example of how science was done in the twentieth century. Extreme physiology was inextricably bound up with political and military motivations and with social and cultural forces; it found ways to create expertise and to control who was and who was not considered an expert; and it relied on the circulation and transformation of data, material culture, ideas, and people in complicated global networks. It is therefore an exemplar twentieth-century science, even though its instruments were carried on the backs of yaks and its practitioners struggled to take down notes with frostbitten fingers. This book is also a continuation of two stories that are currently left dangling in the twentieth century. One, more oriented toward the history of science, is the story of exploration as a form of scientific practice; the other, coming more from the history of medicine, is the story of acclimatization—the biomedical work that tried to make sure that temperate-climate bodies survived in even the most extreme environments.

Exploration as a Way of Knowing

The significance of voyages of exploration for the form and practice of science, natural history, and natural philosophy in Europe has been well established for the early modern period.[2] The influences of these voyages include exposure to new peoples, landscapes, and ideas; the desire to explain new geographical and biological phenomena; the economic rewards available for developments in growing and processing sugar, spices, cinchona tonics, or any of hundreds of other new consumer goods; the ability to collect vast amounts of data or material objects from across the world; and not least, the technical demands for better boat design, rations, and navigation and communication technologies. In the modern period the role of exploration in specific sciences has been clearly argued, particularly natural history and, later, evolution, as exemplified by the largely armchair global botany of Linnaeus and the active exploration of Darwin.[3] Astronomy, too, required its expeditions, including those that sailed in an attempt to record the transits of Venus at the end of the eighteenth century and again at the end of the nineteenth. As geography "militant," in Felix Driver's terminology, established itself in the nineteenth century as a crucial tool of colonial rule, explorers and expeditions were essential to creating the maps and surveys that were supposed to allow the scientific rule of newly conquered empires.[4]

But what of twentieth-century exploration? The twentieth century is, after all, the century in which humans went for the first time to the North Pole (1909 or 1969),[5] the South Pole (1911), and the "third pole" of Everest (1953). Indeed, it is the only century featuring significant exploration of either Antarctica or the areas of the world over 7,000 m above sea level, and it is, of course, the century in which humans left their planet to explore beyond the atmosphere. Although it may be the century when the blank spaces on the map were rapidly filled in, it is also—thanks to improved communications and transport technology, not to mention funding— probably the century in which more expeditions, scientific, military, or leisure, to such spaces took place than in any previous hundred-year period. Historians of geography have made significant contributions to the history of twentieth-century exploration science, as have those influenced by environmental history, especially in oceanography; even so, significant amounts of English-language work on the topic did not appear until the early twenty-first century, notably with *New Spaces of Exploration*, a series of essays edited by Simon Naylor and James Ryan.[6] These works emphasized the continuity between exploration science in the twentieth century and the better-studied explorations of the period before 1900.

In the twentieth century, as in the nineteenth or sixteenth, exploration lays bare the connections between knowledge and power. The desire to "know" a space by mapping it can lead to the claim that one has a right to the space, or owns it. This process can be seen clearly in Antarctica, where the lauded Antarctic Treaty of 1957 gave territorial rights to countries based on their ability to do scientific work there.[7] Exploration also highlights issues of contact and exchange, again as relevant in the twentieth century as in the sixteenth. This book also reinforces the claims of historians of exploration that historical research must go beyond the written word and take seriously the use of images, maps, and other forms of material culture. The following chapters will demonstrate that much knowledge about exploration science was tacit, could be passed on orally and informally rather than through published sources, and in some cases was embodied in physical objects, such as gas masks or abandoned food dumps.

But while the explorations of twentieth-century geographers, oceanographers, meteorologists, and, more recently, biologists and ecologists have attracted historical attention, biomedicine has barely been included in this renaissance of interest in contemporary expeditionary science.[8] In his extensive history of science conducted in the Antarctic, G. E. Fogg states that there was "little that could be dignified by the name of medical research" in expeditions prior to the US Antarctic Service expedition of 1939–41, and even after this point he dedicates only ten pages to medical, physiological, and psychological work in the region.[9] These claims are unfair—as we will see in the chapters that follow, some forms of physiological research were happening in Antarctica long before 1939. Fogg's book does not name the first physiological expedition to Antarctica (INPHEXAN) and skims over the work done by many physiologists whose stories are told more extensively in the following chapters. While the history of altitude science has attracted historical attention (more on this below), the first book-length study of a single high-altitude expedition focusing on its relevance to the history of science and environmental history was published only in 2018—and as it takes as its focus the American Mount Everest Expedition (AMEE) of 1963, physiology is just one among many scientific disciplines being shaped by the challenges of the mountain.[10] It is not just in these extreme environments that medical and physiological field sciences have been overlooked by historians: take, for example, the International Biological Program (IBP)—an enormous international ten-year collaborative project that aimed to do for the organic world what the International Geophysical Year (IGY) had done for the inorganic. While the IBP demonstrated the challenges of doing biology on this scale, and is sometimes regarded as a failure, it involved hundreds of scientists from dozens of countries, funding, supporting, and influencing thousands of research projects between 1964 and 1974. Yet it has only recently received serious attention from historians of science; its most detailed interrogation is by Joanna Radin, who has analyzed the IBP-related work on human evolution and has shown how emphatically racial science was written into the theories and goals, as well as the practices, of the IBP.[11] Assumptions about "primitive" and "untouched" people structured the genetic and evolutionary studies of human bodies, in which the inhabitants of non-temperate zones were expected to act as a window into the evolutionary past of "civilized" White populations.[12] Similar beliefs and patterns of hypotheses can be seen in this book, particularly in chapter 5—and this is no coincidence, as the IBP built on the expeditionary work done, and the theoretical models created, by previous generations. Many of the participants and key actors in the "Human Adaptability" theme of the IBP were researchers who had been involved in extreme physiology work in the 1950s and '60s, and so their pre-IBP work forms part of this book. While Radin has focused on molecular biology, genetics, and studies of heredity, the following pages will add physiology back into that story. It is only a partial recovery, concentrating on the story of survival technologies and acclimatization science, and more work remains to be done on long-term adaptation and evolution, but physiology often forms the bridge between the molecular and the anthropological.

Why should the medical and physiological field sciences have been relatively neglected? I believe there are two intertwined historiographical explanations: the first is the general big-picture story of modern biomedicine that emphasizes the role of the laboratory in the nineteenth century and of "big science" (essentially molecular biology or genetics) in the twentieth. In a historical landscape where the rise of laboratory science dominates and shapes medicine and the life sciences, it is unsurprising that the field as a location of science is marginalized. The second, and perhaps more subtle, explanation is the historiography of field science itself. Historical and philosophical work in the 1980s defined and refined the laboratory as a particular place for scientific work—in its platonic ideal, a "placeless" place, one that, through its rules of entry, regulation, repetition, and a reduction of the natural world to basic principles, could make objective claims to knowledge and truth.[13] While plenty of studies demonstrated that "real-world" laboratories were rather more complicated and messier spaces than this ideal might suggest (particularly in relation to their porous boundaries),[14] this characterization meant that when historians, in this case most notably Robert Kohler and Henrika Kuklick, called for a renewed interest in the *field* rather than the *lab*, the result was a historiography that saw the two sites in opposition to each other.[15] The field was represented as a space whose boundaries were porous, while laboratories, in contrast, were partly defined by their ability to restrict and regulate the materials and people who entered and exited their facilities. The field was also represented as a space whose inhabitants were heterogeneous and often pursuing differing goals—for example, scientists might work side by side with gamekeepers—unlike laboratories, which were supposed to be able to maintain a homogenous and like-minded workforce. And the field was represented as a site in which the scientist had relatively little control or influence over the local environment, most notably in the case of adverse weather, while the most fundamental defining principle of the laboratory was that it provided experimenters with an intensely controlled site in which, at least in theory, single variables could be altered while all the others remained stable. The field, too, often produced knowledge that was localized and specific, while the laboratory's major philosophical claim was its ability to make generalizable, universal, *placeless* claims to truth.[16]

Within this model it can be difficult to take fieldwork as an activity in its own right; rather, it is always a comparator for or a corollary to laboratory work—it is either a form of resistance, rejecting the regulation of the laboratory, or it is a compromise, taking on some of the features of the laboratory in order to make a claim to objective knowledge.[17] Much work so far on the history of twentieth-century medical and life science fieldwork fits into this historiography by categorizing the various sites of non-laboratory work—the "borderlands," as Kohler called them—from the natural or "wild" field site, to the field laboratory, through aquariums and botanical gardens, to laboratories themselves. Alternatively, it may look at techniques, particularly the use of instruments (including the instrumentalizing of large structures, such as research ships) to regulate or control the data produced at field sites.[18] So, while there has been a robust response to Kohler's and Kuklick's call for more studies of the field site, most of the historiography runs in a specific direction, shaped by a particular model of scientific knowledge that opposes laboratory and field and assigns certain behaviors and practices to one or the other—for example, natural history and collecting are of the field, while experiment is the practice of the laboratory.

This book is one of several recent works that challenge this division of practices.[19] While Bruno Strasser and others have shown that "natural history" practices are rife in the laboratory, the chapters that follow show that for many scientists, the wild spaces of the earth were a "natural laboratory"—that is, they provided simplified models, standardized conditions, and unique human experiences that allowed for the production of universalizable knowledge about the body.[20] Indeed, chapter 2 is a case study of a specific situation in which the knowledge produced in the field was repeatedly and robustly demonstrated to be the only source of "truth" about the natural world, as laboratory models and mathematical theories were proved time and again to fail when compared with the data collected on expeditions. The rest of the book tries to take fieldwork, specifically exploration science, out of the shadow of the laboratory and consider it in its own right as a scientific practice. What is most obvious from this approach is how rare it was for any of the biomedical scientists considered here to be only field workers. Almost all of them moved—usually seamlessly—through sites of huge variety, from taking an air sample in a blizzard hundreds of miles from safety to titrating tiny samples of that gas in a warm laboratory on the other side of the world. At times it was a snowed-in Antarctic hut that proved the isolated, controlled space that allowed an experiment to be conducted successfully; at others, the hut's stove leaked carbon monoxide into the air, befuddling scientists so that they were unable to do even basic mathematical calculations.

Other neglected stories are revealed when one moves away from the historiographical constraint that demands a lab versus field divide and which privileges molecular biology as the model science for the twentieth century. The most obvious example is the history of bioprospecting, which is discussed in chapter 4. As a contemporary term, *bioprospecting* means the exploitation of the natural world for economically or socially valuable materials, and it is almost always used to describe genetic or biochemical work, usually on plants, but sometimes using animal or human subjects as its source material. Londa Schiebinger has applied this word to exploration in the early modern period, examining the exploitation of the "natural riches" of empire.[21] Just as historians represent exploration as a core part of scientific practice in the early modern period, but not so much in the twentieth century, Schiebinger's broader understanding of bioprospecting has generally failed to find traction in contemporary histories; most histories of bioprospecting in the twentieth century concentrate on pharmacology, plant extracts, and genetics.[22] Chapter 4 shows how the term can be productively applied to the period after 1900 and to technologies other than pharmaceuticals.

Bioprospecting also reminds us of the centrality of movement and transformation in the use of scientific and practical knowledge. In this book it is not just material objects and behavioral techniques that are transported between the Arctic, Antarctic, high-altitude, desert, and tropical zones, but also living things. For explorers and exploring scientists, the most important of these were people, and their movement is tracked throughout this book. But other animals also mattered, particularly dogs and, early in the twentieth century, ponies, which were moved from Greenland, Siberia, and Alaska to Antarctica for transport, labor, and sometimes food. Other animal transplantations were tried: in 1933 the American Richard Byrd shipped three Guernsey cows to Antarctica, partly as a demonstration that the continent could be made to produce (or at least host) food for settlers and permanent bases, but also as part of a form of celebrity endorsement for Horlicks Malted Milk.[23] In a reverse move, there were early twentieth-century attempts by Norwegian scientists and entrepreneurs to transfer penguins to the coast of Norway (nominally to harvest for their oil, although also potentially for consumption).[24]

Such schemes to transplant economically lucrative animals are also indicative of a second major theme of this book: acclimatization, or the desire to safely relocate living organisms from one ecosystem to another.

Acclimatization, Adaptation, Evolution

Most ancient and early modern theories of health and disease, from Hippocratic humors to miasmas, included an assumption that place mattered to human health and well-being.[25] Not only were some places specifically healthy or unhealthy, but medical theories also frequently asserted that environments had a direct effect on the constitution of an individual, shaping the body and mind and predisposing them to certain sicknesses, personality traits, and futures. Whether these influences were fixed or flexible varied across theories: in some, the locality of birth had permanent, irrevocable effects on the body, while in others these effects could be changed over time or be mitigated by diet and behavior. Likewise, some theories insisted that certain environments were unhealthy for everyone, while others suggested that "healthy" and "unhealthy" were relative terms and that responses to a particular environment were a matter of individual reaction.

The question of the fixity of environmental effects on bodies and health became an issue of political importance in Western Europe throughout the period of exploration and then colonial and imperial expansion beginning around the late fifteenth century. As we saw in the previous section, the movements of material objects and human bodies around a global network were crucially important political activities, and economically vital ones, for most Western European nations by the early eighteenth century. The success of these movements required answers to two questions: could the bodies of Europeans survive in non-temperate climates, and could the economically useful plants and animals of non-temperate countries be successfully grown or bred in Europe (or in colonial territories)? Possibly the most famous attempts to answer the latter question were the acclimatization experiments of Carl Linnaeus, although, as Santiago Aragón and others have suggested, it was the French zoologist Isidore Geoffroy Saint-Hilaire who established acclimatization as a systematic European science.[26] The success of these agricultural and botanical relocation activities varied enormously, but they left clear impacts on the ecosystems of many countries (particularly those involved in major botanical trades: sugar, cotton, tea, coffee)—and Michael Osborne has argued that acclimatization was the paradigmatic colonial science.[27]

By the nineteenth century the responses of humans to shifting environments were a more pressing matter than those of the plants and animals they collected, taxed, or exported from other regions of the world.[28] Permanent White settlement, especially in countries with notably hot, humid, or tropical climates, required a medical approach different from that used for temporary or short-term visitation; in some places, two or even three generations of White families had been born in non-temperate environments, raising new questions about the effect of environment on racial characteristics. The resistance of indigenous people to colonial rule and oppression made the health and vitality of troops, administrators, and settlers a contentious (and possibly financially lucrative) area of biomedical research.[29]

Several historians have already laid out the path that theories about belonging, acclimatization, race, and environment took in Western thought from the late eighteenth to the early twentieth century. As well as Osborne, Warwick Anderson, Mark Harrison, David Livingstone, and Michael Worboys have outlined a clear narrative that takes us to around 1920: initially optimistic, the scientific consensus in Europe toward the end of the eighteenth century was largely that the White body could successfully thrive in non-temperate environments (and likewise that it would be possible to acclimatize tropical plants and animals to temperate environments).[30] This optimism faded through the nineteenth century, partly because of economic and military setbacks, partly because of waves of epidemic disease (such as cholera) that appeared to originate in colonized countries, and partly because of shifting understandings of heredity and evolution. Whether one was a polygenist or a monogenist, the theory of natural selection implied that difference races (species?) of humankind were, to a dramatic extent, fitted to their ecological niches. On the one hand, this theory was used to justify the widespread assumption that tropical races were indolent and uncivilized, and, on the other, it inspired fears that White bodies in tropical environments might be constitutionally unsuited to the climate, or, worst of all, that they might degenerate—they might revert, or decline, to some more primitive state.[31]

Concerns about degeneration in the tropics arose in absolute parallel with fears of degeneration in the industrializing, urbanizing European countries. While it was the "degenerate" urban poor in Europe and the "primitive" colonial subjects who died in the greatest numbers from waves of "Indian" cholera (and other diseases of an apparently tropical origin), the mortality rates of middle- and upper-class colonial administrators began to paint a miserable picture of the long-term future for the White race in the non-temperate regions of the world. Just as critics of rapid, unplanned industrialization and urbanization claimed that city dwellers died out by the third generation and were replaced by healthy rural populations (a pattern that would eventually lead to extinction), claims were made that third-generation White settlers were crippled, infertile, or mentally deficient. While travel guides, colonial doctors, and research physiologists offered advice on survival—which usually involved behavioral changes in settlers' patterns of work, diet, and clothing—the threat posed by the tropical environment to the White body was both short-term sickness and long-term decline. But toward the end of the nineteenth century, biomedicine began to offer a more hopeful account of White survival with the emergence of a new medical specialty: tropical medicine.[32]

The "scramble for Africa" in particular had led to the coining of the phrase "White man's grave" for the western coast of that continent, where a combination of climate, military resistance from local people, and infectious diseases led to extraordinary levels of mortality and morbidity.[33] With the shift from miasmatic to microbial theories about the causes of infectious diseases (that is, their attribution to bacteria, viruses, and other vectors), Western researchers began to offer the possibility of defining a specific cause for the infectious and fever diseases that particularly seemed to affect hot and wet climates. The development first of vaccines and then of chemotherapeutics offered promises of prevention and cure; for diseases such as malaria, the identification of the vector of transmission—often mosquitos—meant that preventive netting and (sometimes environmentally devastating) attempts at extermination could deal effectively with the worst challenges of the non-temperate environment. The work of previous historians has indisputably demonstrated the link between colonial ambitions and the establishment of tropical medicine as a specialty, and as a form of medicine worthy of (state) funding. Indeed, Patrick Manson, the man usually credited as the founder of the discipline, explicitly sold it as a tool of empire: with anti-malarials, preventive vaccination, and chemotherapeutics, White men would no longer die in the tropical grave, but thrive, breed, settle, conquer.[34]

The optimism about the possibility of the "tropical White" lasted a couple of decades into the twentieth century—perhaps most notably in Australia, where some apologists clung to the White settlement of tropical Queensland as evidence that long-term residence was possible for White bodies. This claim was itself part of a deeply racialized rhetoric of displacement and settlement that required the absorption or extinction of Australia's indigenous peoples—identifying them as "hosts" for tropical diseases who needed to be removed from settler environments.[35] But environmental determinism—the idea that human bodies are irrevocably and unalterably "fixed" by their home environment—is a theory that waxes and wanes in popularity. David Livingstone has identified two moments in the twentieth century when belief in this theory was highest, and the first coincides with the fading dreams of the "tropical White" in the 1920s and 1930s.[36] At the same time, forms of tropical medicine began to function as a kind of medical missionary work, no longer primarily supporting immediate conquest and settlement. Instead, international public health and related practices sought to establish Western medical facilities, systems, and patterns of thought in colonized and post-colonial countries.

What has gone missing from this story, as it enters the twentieth century, is short-term adaptation, or acclimatization, science. While for most of the history of Western medicine the environment had been almost indistinguishable from the "disease climate," the microbial revolution at the end of the nineteenth century separated out the effects of (microbial) diseases from the challenges of high and low temperatures, extremes of humidity or barometric pressure, high ultraviolet exposure, and twenty-four-hour darkness. Historians have generally chosen to follow the diseases and the long-term theories about adaptation (e.g., racial science and environmental determinism) rather than the science of short-term acclimatization. There are exceptions—most obviously Warwick Anderson, whose work on physical anthropology and racial science pushes into the early decades of the nineteenth century.[37] While this book cannot extend Anderson's work systematically through the whole of the twentieth century, the intersections between survival science, the physiology of White bodies in non-temperate environments, and racialized science will play out in the background of all the following chapters.

The other major exception to this lack of historical attention is in the specific case of high-altitude adaptation, which was my route into this project. A line in my acknowledgments is not sufficient thanks for Professor John West, not only because his *High Life: A History of Altitude Medicine* is an essential reference point for all work in this area, but also because he worked to create an archive of materials that made it possible for me, and generations of future researchers, to properly explore the topic.[38] Biomedical practitioners and explorers have shown a significant interest in the history of extreme physiology, at least in the case of altitude work, not just by researching and publishing historical work, but also through practices of reliving and memorialization. Their participation reminds us that this is a living science: as I pulled together the first draft of this book in late spring 2017, British researchers announced a breakthrough finding in relation to the genetic basis of the relative immunity to the effects of high altitude demonstrated by Sherpa porters.[39] A century after their superior climbing ability was noticed by Europeans, and sixty years after the first studies suggesting that Sherpa blood responses to altitude were different from those found in European blood, Western science still does not have a complete picture of Sherpa physiology. This book goes some way toward explaining that failure while highlighting the physiological work and discoveries that *were* made through the twentieth century.

Higher and Colder: A Route Map to Expeditionary Physiology

This book is a study of Western science in the long twentieth century, and as such, it has obvious limits. It has drawn overwhelmingly from English-language sources, supported by some in other European languages (and any translations from French or German are my own, unless otherwise indicated); likewise, its examples are drawn mostly from Anglophone or European expeditions and experimental teams. Although this focus is representative of the dominant groups and languages in modern extreme physiology research, it is clear that parallel and alternative work was being done in Russia and China; likewise, a strong indigenous culture of research existed in South America relating particularly to altitude physiology, and I have relied heavily on the work of a series of other historians for the story of Peruvian, Brazilian, Andean, and Mexican research, including Marcos Cueto, Jorge Lossio, Stefan Pohl-Valero, and others.[40]

This is also a deeply masculine story. It is hardly a dramatic finding that men and men's bodies dominated an activity involving scientific work and exploration, both practices heavily coded as male in the twentieth century. As we shall see in chapter 3, those who controlled access to extreme environments used both "soft" and "hard" bans to exclude women, making them a tiny minority as either subjects or practitioners until well into the 1980s. That said, it has been possible to recover women's participation in new ways. During the course of writing this book, the #thanksfortyping hashtag became popular on Twitter—a way of commenting on the tendency for male authors and researchers to reduce or minimize the roles of women (often wives or junior scholars) in academic and research work. In the acknowledgments sections of articles, books, even grants, I did begin to find women: they were typing, they were packing boxes, and they were funding expeditions. Pioneering historical work is finally being done to recover the stories of women who worked in extreme environments—and this book should emphasize the fact that their stories do exist, but that more effort and more imaginative use of sources will be required if historians are going to rediscover them.[41] Just as one example, the published materials relating to the Silver Hut expedition (a physiological expedition to high altitude in the mid-1960s) mention that the scientists' wives came to visit them at Base Camp, and the archival materials add that there were personality clashes and tensions caused by their visit, but it was only when I went through some photographs relating to the expedition that I spotted one showing a woman holding a meteorological instrument and another of a woman apparently assisting with experiments by taking notes—participation that was invisible in the printed sources.[42]

In many cases women worked as essential research partners of male explorers, yet they are almost erased in the scientific representation of that work. Another clear example is Joan Rodahl, whose husband Kåre Rodahl became one of the world's foremost experts on Arctic life and the "Eskimo" in the mid-twentieth century. (A note on terminology here: the word *Eskimo* is widely used by my sources throughout the period considered here, while the alternative term *Inuit*—introduced around 1977—is still rejected by some circumpolar peoples; generally in this text I will be using all three terms as they are used in my original sources.)[43] Jean Rodahl traveled with Kåre and was frequently responsible for the packing and logistics of his extensive medical and survival equipment, assisted in his medical clinics, took down notes and figures and measurements, and provided assistance to such a level that, were she a male medical student, she could have expected co-authorship on papers. But when later explorers referred to "Rodahl's Eskimo," they meant the singular—the Eskimo as described, studied, and "discovered" by Kåre—rather than "the Rodahls' Eskimo," representing the work done by a couple.

That misplaced apostrophe might seem like obscure and subtle typography, but it is in such subtleties that women are often discovered. For some time I failed to realize that one of the major mid-twentieth-century studies of cold adaptation in man had been largely analyzed and written up by a woman—Mrs. R. J. Sutherland—simply because the report she wrote eschewed the traditional practice of flagging female authors by using "Miss" or "Mrs." and instead just listed her by her initials and surname (more on Sutherland in chapter 3). Likewise, for some time I insisted that no women were doing physiological or biomedical work in the high Himalaya in the 1950s, only to discover two of them toward the end of the project: Nea Morin, a pioneering female climber who contributed to ECG studies, and a female member of the London School of Economics' Mountaineering Club, who was a subject of (and possibly a researcher in) sleep studies. This book's structure therefore reflects the reality of women's participation in this work—there is no ". . . and women" chapter collating their activity; instead, they are threaded throughout, reflecting the reality that while they were marginal, they were present everywhere.

The role of non-White participants in this research is also often obscured by the conventions of scientific publishing. Although any paper based on fieldwork in the high Himalaya is dependent on the work of Sherpa and other porters, they rarely feature in the acknowledgments, functioning rather as "invisible technicians" whose physical (and sometimes mental) labor is taken for granted.[44] There are some exceptions, notably for elite or European-educated non-White participants. International research teams, particularly in the high Himalaya, occasionally included non-White researchers— the most high-profile perhaps being Dr. Sukhamay Lahiri, who was born in what is now Bangladesh, educated at Calcutta and Oxford Universities, and who went to the US on a Fulbright Scholarship in 1965 and remained there until his death in 2009. Lahiri worked extensively on altitude physiology; his "breakthrough" expedition was the Silver Hut expedition of 1960–61 (see chapter 2). In South America, scientists from Europe and North America often collaborated with local researchers (and businesses) in order to get access to facilities and human subjects at mid-altitude sites—indeed, the question of high-altitude physiological adaptation performed a role in creating and supporting (politically useful) national and ethnic identities. This interaction between nation building, identity creation, ethnicity, and physiology in South America is better outlined by other authors, but is considered here in some detail relating to the work of the Peruvian scientist and high-altitude physiologist Carlos Monge Medrano.

Inevitably, it is the voices of the elite participants, particularly the scientists, that are easiest to recover: it is difficult to know what the Andean miners of the 1930s thought of the American and European (or Peruvian) scientists who turned up at their workplaces to take their blood and measure their chests. Sherpa did not become the subjects of physiological studies in any significant numbers until the end of the twentieth century, and when discussed in relation to the *practice* of scientific work, they are more often cited as aggravations—stealing or breaking technology—than as essential helpers. Much of my discussion of Sherpa participation is informed by the seminal anthropological work of Sherry Ortner, who clearly demonstrates how two cultures—Sherpa and Sahib—functioned on the mountain, and who undermines the representations of the former by the latter, providing a much more nuanced understanding of the relationship between European climbers and local guides.[45]

The Antarctic, without an indigenous population, is a white space in more senses than one. And while "Eskimo" are less used as guides in the Arctic than Sherpa are in the Himalaya, they can sometimes be discovered as assistants and participants in expeditionary science (as a researcher in the 1970s wrote, "Diary-keeping then became a problem due to . . . loss of hand-coordination; in some instances, it was thus necessary to use a well-acclimatized Eskimo assistant as a temporary scribe").[46] The various indigenous peoples of the circumpolar region were also scrutinized, as part of the IBP and other surveys, by physiologists, anthropologists, survival scientists, and geneticists in the hope that their bodies and cultures would provide insights into White survival in extreme environments. Wherever possible, this book has tried to amplify the voices of marginalized groups—in particular, to acknowledge that scientific work depends on a complex collective of dozens of people, including equipment designers, assistants, transport technicians, guides, human computers, and typists. But the nature of the subject—the colonial activity of exploration, the masculinized activity of science—and the limits of my language skills mean that this history remains a White, male, Eurocentric one.

It is also the history of a very modern category, albeit an actors' category. The phrase *extreme physiology* was rarely used in the twentieth century, except to refer to the "extreme physiology" of extremophile animals.[47] It has been increasingly used by physiologists in the twenty-first century as a term for the physiological study of human limits and of the human body in extreme environmental conditions—as codified by the founding of the journal *Extreme Physiology & Medicine* in 2012—and that is the sense in which I am using it in this book. Some other words used in this book have more ambiguous meanings, and perhaps most notable here is the extremely confusing slippage between the terms *acclimatization*, *adaptation*, and *acclimation*. As a general rule, in this text, I use *acclimatization* to indicate a short-term accommodation to a new environment or a medium-term one achieved within the lifetime of an individual, while *adaptation* implies a longer-term, probably hereditary change in function—although the scientists, explorers, and historians quoted in this book have tended to use the words more interchangeably, so they should always be understood in context and not as absolutes. *Acclimate* and its derivations, although used in French writing, is a much less common variant in the twentieth century (although it has become more popular in the twenty-first), and I have avoided its use to try to prevent an unnecessary multiplication of terms. Finally, the names of people and places here are given in their contemporary Anglicized form in my text, but I have left original spellings in quotations where relevant; therefore, for example, Solukhumbu may be rendered as Sola Khumbu.

Chapter Outline

Chapter 2, "Gasping Lungs," starts at the chronological beginning of this book's story, with mid-nineteenth-century experimentation on low barometric pressure, before going through the twentieth century's most influential investigations into altitude physiology. The question of what would happen to a human body on the summit of Mount Everest is an excellent case study of the conflict between laboratory models and real-world experience; time and again, respiration physiology has been a science in which the truth of the field contradicts the hypothesis of the lab bench, the barometric chamber, or the mathematical model. The story of altitude physiology is also one that demonstrates how other historiographies and historical mythologies can shape and obscure scientific practice. In this case the strong and abiding appeal of the myth of British amateurism (or, as it is sometimes phrased, *gentlemanly* amateurism) has led to the neglect, or even unfair representation of, the dedicated and intense scientific work that backed up the first forty years of attempts on Everest.[48]

The representation of science done in the Antarctic has also been affected by traditional histories and mythologies, particularly by debates over leadership styles, heroism, and blame seeking in biographical work on Robert Falcon Scott, Roald Amundsen, Ernest Shackleton, and others.[49] Chapter 3, "Frozen Fields," shifts the story to Antarctica and considers the ways in which small exclusive communities of highly networked scientists and explorers formed around the scientific work done in extreme environments. While not entirely consisting of "gentlemen" (and containing almost no "amateurs," as experience was a crucial quality sought in expedition members), these cliques rarely contained "ladies"; the inaccessibility of their research sites, and the unwillingness of their members to expand their circles, excluded women—and indigenous people—from participation until the very end of the twentieth century. The creation of these fields of expertise, and the networks that sustained them, was achieved through a broad range of activities, many of them outside the traditional systems by which expertise and authority are created or advertised in the sciences. While participants did publish papers, host conferences, and engage in the more traditional practices of esteem and reputation building in the sciences, personal connections were crucially important to the selection of research teams. Information was frequently passed through and around international networks (a true Republic of Letters for the modern age), often in the form of material culture—examples of equipment, blood samples, or food rations—rather than as text. Networks, even important emotional connections, could be created without the physical meeting of human beings, as the chapter demonstrates in the case of materials left behind in dumps or caches for friends or strangers on mountainsides and in the polar regions.

Transformations of material culture, particularly transformations of indigenous knowledge into Western scientific technology, are the core subject of chapter 4, "Local Knowledge." This chapter brings the Arctic into the story and considers the many varieties of local expertise of value to physiological expeditions and biomedical exploration. The chapter divides local knowledge into three categories: embodied knowledge, meaning biological or physiological adaptations to local conditions, usually considered a racial or ethnic characteristic; environmental knowledge, meaning localized information about defined regions or specific conditions (e.g., the locations of water sources or warning signs of avalanches); and survival technologies, meaning the material objects and practices of survival that are universalizable—that characteristic so strongly associated with laboratory knowledge—and can be transported from one region or expedition to another. The chapter shows that "local" knowledge could be transformed into "general" expertise, but that assumptions about the portability of knowledge were often based on whether that knowledge originated with Western explorers or local peoples. Some of that knowledge could be bioprospected; that is, taken, transported, and often transformed into various kinds of Western scientific knowledge. The chapter also looks closely at quotidian technologies, such as housing, clothing, and food—none as glamorous as high-altitude oxygen systems, but all as crucial in terms of survival and as essential to the success of scientific expeditions. Vigorous attempts were made to render scientific the studies of complicated subjects such as taste and nutrition or thermal stress and comfort, including the use of new measuring instruments and units; in many cases researchers found themselves emphasizing the complex and individual nature of these bodily experiences and their resistance to standardization and quantification. The expertise of exploring scientists, so important to their entry into, and status in, the networks described in the previous chapter, could easily be embodied in their varieties of local knowledge—at least of the environmental and technological kind, since embodied knowledge proved more problematic, as the subsequent chapter will demonstrate.

Chapter 5, "Blood on the Mountain," which returns to questions of acclimatization and adaptation, roams across all the spaces considered in the book, adding the deserts of Australia to the Arctic, Antarctic, and mid- and high altitudes. Specifically, it looks at the relationships between the physiological study of the White and the non-White body; between indigenous people and Western visitors; and between theories of short-term acclimatization and long-term evolutionary explanations for racial and ethnic differences. It also returns to the question of the second chapter—the difference between laboratory and natural models—in considering studies into the possibility of so-called artificial acclimatization. In addition to these binaries of White/non-White and natural/artificial, it considers the ways in which women's bodies were used, in contrast to men's, in physiological research.

Following on from the explicitly racialized, and likewise misogynistic, discussions and assumptions built into some extreme physiology work in the twentieth century, chapter 6, "Death and Other Frontiers," concludes the book by considering the broader ethical challenges of experimentation in expeditionary biomedical science in the long twentieth century. As well as relying on (often working-class) indigenous peoples, experimental extreme physiology required the willing labor of (often elite) Western explorers, scientists, and sometimes athletes; the enforced participation of military recruits; and sometimes the unknowing participation of the dead. This chapter also reviews the findings of extreme physiology through the twentieth century, showing how its discoveries and its uncertainties shaped research that has affected everyone from premature babies to Sunday hill walkers.

Much of this book demonstrates that extreme physiology is a scientific activity that, as a model, tells us a great deal about the way modern biomedicine is practiced: it is done through large, internationally connected networks of support and exclusivity; it is done through the sharing of material culture, samples, informal advice, and personal embodied experience as well as published materials; it is both constrained and liberated by complex multisource funding models; it enables knowledge and technology to be moved around the globe while sometimes effacing, sometimes emphasizing, and sometimes inventing its origins; it is done by scientists who move between research sites, balancing the needs of laboratories and field sites; and it can be done only with the assistance of a vast, usually unseen and uncredited, army of invisible technicians—from porters to instrument makers to wives. At the same time, as chapter 6 will emphasize, it is also an extraordinary and atypical science that has made extreme physical demands on the bodies of its participants, and in many cases has killed them.

TWO

Gasping Lungs

On March 30, 1874, the Frenchman Paul Bert recorded in his diary a success-ful climb to the summit of Mount Everest.[1] His only living companions on the ascent were a rat and a sparrow in a wicker cage, who both struggled in the thin air; this was not a problem for Bert, who was connected to a system supplying him with oxygen. Bert examined the sparrow carefully, as it had become distressed and vomited on the ascent. He increased its discomfort by inserting a rectal thermometer and recording that the unfortunate bird's body temperature had fallen over 5°C compared with the measurement he had made at sea level. The candle Bert had brought with him eventually flickered and went out, but despite this setback he wrote that he would like to have gone higher still, but unfortunately his steam pump broke.

Bert's story will come as a surprise to those who know 1953 as the year when the summit of Everest was first reached by climbers. Bert's "ascent," less than twenty years after Everest was recorded as the world's highest mountain, actually took place nearly 4,000 miles away from the Himalaya, in a baromet-ric chamber installed in a research laboratory in the Sorbonne, Paris. None-theless, Bert was convinced that his ascent proved that even the highest of the earth's mountains was not, in theory, inaccessible to man ("du mont Everest, la plus élevée des montagnes du globe [8,840 m], n'est plus théoretiquement inaccessible à homme").[2] The use of Everest is in part a familiar compara-tor, added to Bert's book, *La pression barométrique: Recherches de physiologie expérimentale*, to convert an abstract measurement (height in meters) into a real-world example (a particular mountain).[3] We still often use geographical objects, most defined by human boundaries, for this purpose: things are as big as Wales or Belgium or Texas, or as long as the Amazon. At the same time there is a real claim of equivalence being made here: the work Bert was do-ing in his laboratory was producing factual understanding of the reaction of human (and sparrow and rat) bodies on the summit of the world's highest mountain. It took another 75 years for a human body to arrive on the actual summit, and 107 years for a team to take clinical measurements—although not rectal temperatures—near the summit. We had to wait 133 years from Bert's experiments until women's bodies were studied on, or near, the summit of Everest. We are still in the dark about sparrows and rats.

What becomes clear when we look at those thirteen decades of research—which will be sketched out in this chapter—is that Bert's study is something of an anomaly. Time and again, from the late nineteenth to the early twenty-first century, the most authoritative scientific knowledge about bodies on Everest has come from studies on mountains, not in chambers. This is not to say that laboratory work was neglected, nor that publications based on mountainside research outnumber those created with barometric chambers and lab bench equipment. On the contrary, this chapter will show that mountaineering expeditions to high altitude were often tightly connected to networks of research that included laboratories—in universities, in military establishments, in industrial research and development departments—as well as other field sites across the globe.

Thousands of expeditions to high altitude have been made since the beginning of the twentieth century. Here we concentrate on a relatively small sample of these expeditions, including a few general national expeditions as well as specifically scientific ones. They have been chosen because each significantly contributed to answering the question Bert implicitly posed: Could humans climb to (and survive on) the summit of the world's highest mountain? Over a century, the question evolved from Bert's original investigation into the cause of physiological disturbance at altitude, to attempts to find practical technological cures for that disturbance, to detailed studies of the body's natural responses to that disturbance and the possibility of permanent biological adaptation to high-altitude environments.

Through this evolution, a deceptively simple question in physiology leads to a demonstration of the ways in which the field and the laboratory may conflict—and in this case it is the field that ends up as the ultimate arbiter of scientific facts, despite the risks and challenges a site such as the summit of Everest poses to scientists and their human subjects. These challenges are literal and physical and sometimes fatal, and they are also political and social, as the mountain (and others like it) has been a site of international conflict and competition. In particular, the risk of a "scramble" for Everest in the 1950s renewed British interest in what had sometimes been a desperate attempt to claim an extraordinary physical feat and an exploratory first for a fading post-imperial nation. By the end of the chapter, it will be obvious that

Bert's research is different in tone and approach from the general thrust of the altitude research that followed. In some ways he is not atypical: the majority of researchers who have worked on issues related to altitude probably have never climbed any mountain, let alone Everest. Studies of blood and air samples, and the use of machines, mathematical models, and simulators as well as animals as proxies for human bodies, remain core work in this research area. But Bert's confidence in his approach to solving the problem that was altitude sickness, his faith in the fit between laboratory model and real world, is a candle that flickers—and is sometimes extinguished—through the following century and a half.

Climbing Headache Mountain:
The Problem of Altitude Sickness

By the time Bert climbed into his barometric chamber, the fact that humans experience strange bodily symptoms on high mountains had been known for centuries. Bert's book, *La Pression*, starts with a long chapter of collected stories about altitude sickness, from early accounts originating in China around 30 BCE, through work by Humboldt in South America, to the most up-to-date observations from high-altitude balloonists. A coherent syndrome, or collection of symptoms and experiences, emerges from these reports: headache, shortness of breath, fatigue, disturbed sleep. Like any disease, what became known as "mountain sickness" seems to vary among individuals, so the symptoms come on at different heights for different climbers (and much faster for balloonists), and some experience additional or alternative effects such as vomiting, loss of appetite, or mental disturbance. The severity of the symptoms differs not only among individuals, but even from mountain to mountain. Bert collects these stories for two reasons: first, they are his "route to the real"; that is, he is using the stories to illustrate and prove the existence of a real-world phenomenon, which he is then going to analyze and understand in the laboratory. Second, and just as importantly, the anthology of mountaineering stories also gives him an opportunity to list, and almost entirely dismiss, most of the existing explanations for altitude sickness. The most popular of these explanations were that the sickness was caused by a poisonous gas or wind, or by the physical derangement of internal organs, due to either muscle fatigue or changes in air pressure.

What Bert did not dismiss were the theories of his colleague, the doctor and physiologist Denis Jourdanet. Jourdanet had spent nearly eighteen years in Mexico between 1841 and 1869, interrupted only by time spent in France completing his medical studies.[4] He had published a series of works on the effect of altitude on health and disease, possibly the most important of which for Bert was his 1863 paper "De l'anémie des latitudes et de l'anémie en général dans ses rapports avec la pression de l'atmosphère," in which he makes a clear case for the "anaemia of altitude," suggesting that altitude sickness is caused by hypoxia and citing his laboratory and field observations showing that humans at altitude have lower arterial oxygen concentrations than those at sea level.[5] It was also Jourdanet who first suggested that Bert, as a young researcher, might turn his interest in respiration and anoxia to profitable study by considering the problem of altitude sickness. Perhaps even more importantly, it was Jourdanet who provided the funds for him to do so.[6]

Bert's approach to research was fundamentally shaped by the French physiologist Claude Bernard, who championed the use of laboratories and animal models in the sciences related to medicine. More than just a practical method of work, this approach also came to symbolize an intellectual and philosophical attitude toward making scientific facts. Although Bernard recognized that living things were holistic systems, intimately connected and infinitely complex, his solution to this problem was a reductive approach to experimentation: to selectively disrupt or destroy particular parts of the system to study their effects on the whole. By attempting to isolate and manipulate individual organs, chemical balances, and inputs and outputs, his methodology offered the promise of creating knowable order out of the chaos of biological complexity.[7] This, essentially, is the approach that Bert took: out of the complex of symptoms that made up the disease (or possibly diseases) that constituted mountain sickness, Bert refined his research question to one specific syndrome, *altitude* sickness. Further, he started with the assumption that the explanation for this sickness could be found in the intersection of two body systems—the respiratory and the circulatory. Following from Jourdanet, his working hypothesis was that altitude sickness (and by inference, mountain sickness) was a form of hypoxia, caused by the reduced partial pressure of oxygen (P_{O_2}) experienced at altitude.

Bert's research program was a detailed one, and *La Pression* is the summation of about a decade's worth of work—crammed in between the other commitments required of the chair in physiology for the Sorbonne and his political activities (he took a seat on the Paris Assembly in 1874 and was elected to the Chamber of Deputies in 1876).[8] This program of work, which reaches an apogee in his barometric chamber experiences, can be simply summed up: a lack of oxygen causes the symptoms associated with mountain sickness, and supplementation with oxygen alleviates those symptoms, in rats and birds as well as in humans. From a medical standpoint, then, mountain sickness is revealed as altitude sickness, a simple imbalance in the *milieu intérieur* caused by a lack of oxygen; and the results in the laboratory unproblematically show that with suitable supplementation, the highest peaks in the world are attainable by man (and rat and bird).

While this finding supported Jourdanet's arguments, other researchers were not convinced. The most rapid critique came from Marc Dufour, the president of the Société Médicale de la Suisse Romande. A specialist in diseases of the eye, Dufour had written his thesis on force and fatigue in muscle, and this work had led him to the conclusion that fatigue played a role in mountain sickness—perhaps as an explanation of the fact that the height at which climbers experience the symptoms seems so variable.[9] Dufour pointed out that, far from showing a systematic relationship between a single variable—oxygen—and a set of symptoms, the disease presented with more variation than Bert's reductive explanation could predict. While Dufour did not deny the role oxygen played in producing symptoms, he suggested, after reading Bert's work, that it may act in conjunction with fatigue, so that a tired mountaineer might experience symptoms at a lower altitude (and therefore a higher P_{O_2}) than a fresh and fit mountaineer.

Dufour certainly had a point: the crux of Bert's research was to study only one factor, one disruption to the *milieu intérieur*, at a time. However convincing the results were within the contained sphere of the laboratory or the chamber, there was no guarantee that, when additional complications were experienced on the mountainside, the "simple" solution of oxygen supplementation would continue to work. Alpine climbers already knew the problems of tiring ascents, freezing temperatures, high winds, and mental fatigue and stress; as Western explorers entered the Himalaya, their initial high-altitude climbs made the difference between the laboratory and the mountain even starker. Bert's "ascent" of Everest did not include weeks of trekking to Base Camp, temperatures as low as $-60°$C ($-76°$F), bouts of dysentery, or the loss of companions to avalanche and crevasse. It was therefore an obvious critique of Bert's methodology that his "ascent" of the mountain was experienced while sitting in a chair and that the most energetic activity he engaged in was the inserting of thermometers into the rectums of hypoxic birds. Later experimenters, with larger chambers, added exercise bicycles, treadmills, and step boxes in an attempt to create fatigue. Even these technologies did not address the possibility that *mental* as well as physical fatigue had a significant effect on the experience of mountain sickness.[10]

Fatigue itself remained an opaque biological process well into the twentieth century, and its study is closely associated with work on the *milieu intérieur*. In the twentieth century this concept was reinterpreted as *homeostasis*, a word coined and popularized in the 1930s by physiologist Walter Cannon.[11]

Cannon, as professor of physiology at Harvard Medical School, was a key player in the founding of the Harvard Fatigue Laboratory, an institution that not only revolutionized our understanding of fatigue but also studied altitude and other topics in extreme physiology.[12] Researchers at the Harvard Fatigue Laboratory conducted expeditions to mid- and high-altitude sites in both North and South America.[13] Although they championed a reductive approach to physiology, studying systems in their component parts, they also recognized that altitude sickness, like fatigue, was a disorder of the whole body, a disruption in the homeostasis of a living organism that affected all body systems. Unlike Bert's work, their laboratory studies were explicitly conducted in connection with expensive and extensive fieldwork.

As such, it is not Bert's approach, but that of later physiologists, that best epitomizes the study of altitude physiology that emerged in the twentieth century. As a founder figure, we might instead turn to Angelo Mosso,[14] an Italian physiologist of eclectic interests who had worked for a time in the laboratory of Claude Bernard. Mosso's research into altitude physiology started in earnest when he began to use the newly constructed laboratory hut on Monte Rosa, one of the Italian Alps.[15] Inspired by suggestions that a high-altitude observatory ought to be built in the region, the Capanna (Regina) Margherita was a fairly humble space divided into living quarters, a kitchen, and a laboratory, which was extended in 1898 to increase the space for research (see figs. 1 and 2). After several renovations, it remains the highest research station in Europe to this day, at 4,450 m above sea level.[16]

The Margherita laboratory extension is sometimes referred to as an "observatory"—Mosso used the phrase "Monte Rosa Observatory" to describe the hut while it was being planned—and it has been used for astronomical and meteorological research.[17] By the 1890s European astronomers had recognized the value of studying the night sky from relatively isolated outposts on mountainsides. "Relatively" is an important modifier here, because although these observatories performed an important function in removing the researcher from the literal and moral hubbub of the growing industrial cities of Europe, they were also far from isolated. In fact, the sites were usually chosen because they were well networked—connected to the rest of the world via telegraph, traditional walking routes, established portering arrangements, even railway lines.[18] Most of these mid-altitude research sites therefore had mixed uses: on the Faulhorn in the Swiss Alps, a hotel also functioned as a base for physiological and astronomical research. Similarly, a hut like the Capanna Margherita, designed for physiology, could also be used as a space for physicists, astronomers, doctors, and exhausted travelers.[19]

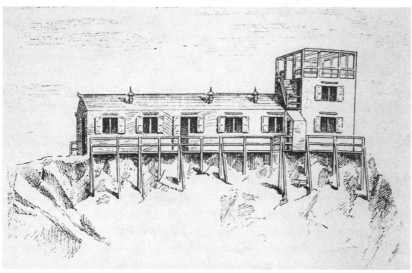

1. The original "Margherita hut" (*above*) and a sketch of the extended Capanna Margherita (*below*). (Both taken from Mosso, *Der Mensch Auf den Hochhalpen* [1899], pp. 164, 428.)

Mosso fits perfectly into this pattern of mixed use because his own researches were part of a network that included diverse and eclectic interests. As well as problems of altitude sickness and fatigue—which Mosso saw, like Dufour, as intimately connected—he also studied nutrition, wrote about emergency medical procedures on the mountain, considered how best to breed animals at altitude, and conducted pioneering dermatological studies: concerned about sunburn, he tested turmeric, red ochre, black lead, soot, Vaseline, and grease as skin protectors—sometimes applying a different paste to each half of his or a colleague's face, or one color to the nose and another across the cheeks. This "kept us in a cheerful mood and proved a great source of merriment to the mountaineering parties with whom we came into contact."[20]

When it came to the problem of altitude sickness, Mosso used his experience on the mountainside to directly contradict the conclusions of laboratory-bound Paul Bert. He challenged claims, repeated by Bert although originating with Horace Bénédict de Saussure, that people breathe more rapidly at altitude, suggesting that studies had been done only on people still exhausted from a climb, not climbers at rest. This mattered, as he reasoned that hypoxemia (low blood oxygen), which Bert thought occurred due to low P_{O_2}, would lead to an *increase*, not a decrease, in respiration.[21] Mosso also pointed out that asphyxia—that is, a lack of oxygen—is usually associated with a decrease in the pulse rate, but at altitude the pulse is increased.[22] Further, Mosso pointed to empirical and practical evidence that oxygen is not a cure-all for altitude sickness. On a climb of Mont Blanc, he says, he "learnt one thing" from talking to the guides and workmen who had built the mountain observatory, which was that "oxygen is of no use against mountain-sickness."[23] More dramatically, he relates the case of Dr. Jacottet, a young French man who died suddenly on Mont Blanc in 1891. Jacottet experienced serious symptoms of breathlessness and fatigue, and although advised to descend, he refused. He wrote to a friend saying that although he was suffering from mountain sickness, he was going to stay in order to study "the influence of atmospheric depression and acclimatise himself."[24] The next day he suffered fits and apparent paralysis, and although oxygen was given to him, he died early the following morning.

Therefore, the *theoreticament* of Bert needed to be tested against the *réalité* of the mountain. Mosso did not downplay the value of laboratory studies—indeed, he moved almost seamlessly between sea-level laboratories, sea-level field studies, laboratories on the mountain, and mountaineering expeditions (as well as incorporating a vast secondary literature) while consolidating his evidence on the effects of altitude. But unlike Bert, Mosso was trying to describe the holism, not find its crucial weak point—as evidenced by the title of his key publication in this area, *Life of Man on the High Alps*. The word *life* in his title accurately indicates his aim, which was to describe the effects of mountain living and mountain travel in all its aspects. Specifically, however, *Life of Man* suggests that it is carbon dioxide, not oxygen, that is the important gas in the production of altitude sickness. Ralph Kellogg, an American physiologist with an interest in history, has carefully gone over Mosso's figures and has argued that his interpretation may be due to a mathematical error (converting gas analyses to percentage by weight instead of by volume) and a practical mistake (using a barometric chamber that was too small for the purpose); it is interesting to see that even in the view of modern commentators, it is the *laboratory* that has led Mosso astray, rather than the field site.[25]

Although contemporary researchers certainly look back to Bert as the founder of their field, it is clear that at the end of the nineteenth century no consensus existed about the cause (or the cure) of altitude sickness.[26] In 1906 Dr. T. G. Longstaff, both a qualified doctor and a mountaineer, and later a member of the British Mount Everest expeditions of the 1920s and '30s,

2. The crowded experimental conditions inside the Capanna Margherita (original caption: "Von meinem Bruder in der Hütte Königin Margerita [*sic*] [4,560 m] angestellter Versuch, um die Menge der während einer halben Stunde eliminierten Kohlensäure zu bestimmen"). (Taken from Mosso, *Der Mensch Auf den Hochhalpen* [1899], fig. 49.)

published *Mountain Sickness and Its Probable Causes*, in which he argued that only fatigue and personal susceptibility could explain individuals' varied experiences of altitude sickness.[27] Dr. Fillippo di Fillippi, the medical officer for an Italian explorer, the Duke of the Abruzzi, also implicated fatigue when he reported in 1909 on the lack of sickness among the Duke's high-altitude team.[28] So while it is Bert's conclusion, and not Mosso's, that is now considered the "correct" interpretation—that is, low P_{O_2} is now considered the proximate cause of altitude sickness—Bert's pure, laboratory-only, single-factor studies are a relatively poor model for the altitude research that followed. Although this chapter does not have the capacity to chart all the research conducted into questions of altitude, anoxia, and the human body, the studies it discusses are—at least in hindsight—the most significant, the ones that kick-started careers and lifelong research interests, that emanated from or created the world-leading research centers. That virtually all of the studies considered here are expeditionary experiments is not the predetermined bias of this book, but rather a genuine representation of the balance of power between laboratory and expedition: laboratories provided groundwork, data crunching, and mathematical models, but time after time it was the mountain and the mountaineer that confirmed, denied, or created new facts about the human body.

Another feature of Mosso's work that makes him a good role model for the scientists who followed is his dense network of collaborators and co-researchers. Mosso spent time with or visited the laboratories of many of the leading international figures in physiology, including Claude Bernard and Étienne Marey in France and Carl Ludwig in Germany, while his own laboratory (est. 1879) in turn became a hub for visiting physiologists, including the British researchers Charles Sherrington and Harvey Cushing. Of all his intellectual collaborators, it is Nathan Zuntz who deserves the most attention: a chemist and physiologist, Zuntz conducted studies at the Capanna Margherita, along with Adolf Loewy and others, into the consumption of oxygen at rest and at work at altitude, into changes to blood composition, and quite extensively into diet.[29] Zuntz's research also spanned field and laboratory, as he studied altitude in the barometric chambers of the Jewish Hospital in Berlin in the 1890s as well as on Monte Rosa, turned his attention to ballooning in the early years of the twentieth century, and then organized the International High-Altitude Expedition to Tenerife in 1910.[30] The Tenerife expedition—which included two British researchers, C. G. Douglas and Joseph Barcroft—conducted blood and respiratory studies at 2,130 m and 3,350 m above sea level.[31] Here, Barcroft appeared to disprove Mosso's assertions about altitude sickness—namely, that it was caused by lowered partial pressure of carbon dioxide (P_{CO_2}) in the blood rather than, as Bert would have argued, lowered P_{O_2}. Barcroft was the only member of the team to suffer significantly from altitude sickness, and the only one not to experience a significant fall in his alveolar P_{CO_2}, which remained largely unaltered by the ascent.[32]

The International High-Altitude Expedition to Tenerife was the first European expeditionary experiment of the twentieth century, and perhaps one of its most important outcomes was that it acted as an encouragement to further—and higher—mountaineering research. In 1911 Barcroft took an expedition to the Capanna Margherita, and Douglas co-organized a more adventurous trip overseas, to Pikes Peak in Colorado, USA. The Anglo-American Pikes Peak Expedition of 1911 included Douglas and J. S. Haldane along with two Yale-based physiologists, Yandell Henderson and Edward C. Schneider. In Colorado they were joined, extremely unusually for these sorts of expeditions, by a woman, Mabel Purefoy FitzGerald, who had studied physiology at Oxford University, had worked with Haldane in his laboratory, and was in New York studying bacteriology on a Rockefeller Traveling Fellowship. FitzGerald was not invited to take part in the laboratory work on Pikes Peak and instead (in collaboration with Haldane) devised an independent program of study, visiting mining camps in Colorado to study the blood P_{CO_2} of acclimatized miners.[33]

Interestingly, among the justifications given for the choice of Pikes Peak in 1911 is the suggestion that working on a mountainside allows for "the physiological conditions, apart from the reduced atmospheric pressure . . . [to be] normal as far as possible."[34] That is, the "natural laboratory" of the mountain allows for a single environmental factor, altitude, to be adjusted while the others (fatigue, diet, personal space, and so on) remain "constant," thus figuring the mountain as a more realistic and controlled site than the barometric chamber.[35] Logistically, Pikes Peak was a more favorable research site than alternatives in the Alps, Andes, or Himalaya because of the availability of accommodations and a cog railway—important not just for supplies, but also because it ferried up unacclimatized tourists who could be observed as comparators for the acclimatized experimental subjects. The confident conclusion of the expedition was that Bert, not Mosso, had identified the key cause of altitude sickness: hypoxia. These researchers also, like Bert, extrapolated from their experiments to much higher altitudes, suggesting that

> it would be absolutely hopeless for an unacclimatised person to climb to the height of Mount Everest, or to reach it by aeroplane or dirigible balloon without the use of oxygen; but there seems to be no physiological reason why an acclimatised person should not succeed in climbing it, dangerous as the attempt might be.[36]

Only Rotters Would Use Oxygen?

Although the expeditions described so far were at altitude, they were still some way short of generating, in the field, a knowledge of the human body at the heights Bert had studied. It was not until the 1920s, when the British turned their eyes to the "third pole" of Everest (having failed to be the first to reach either the North or the South Pole), that the possibility of putting, and studying, a human on Everest began to be seriously considered. The British mounted seven expeditions—five full attempts on the summit (1922, 1924, 1933, 1936, 1938) and two reconnaissance expeditions (1921, 1935)—in fourteen years. Until recently, the story of these expeditions pitted an amateur gentlemanliness against scientific "progress," blaming the failure to develop (or in some cases, use) effective oxygen systems on the attitudes of traditional mountaineers, who dismissed such technical assistance as "cheating." This is certainly the narrative provided by the account of the successful British-organized attempt on Everest in 1953, in which the physiologist Lewis Griffith Cresswell Evans ("Griff") Pugh and the surgeon (and team doctor) Michael Ward blame "the futile controversy over the ethics of using oxygen, and the failure to accept the findings of pioneers in its application, [which] handicapped for thirty years the introduction of a method which promises to revolutionize high-altitude mountaineering."[37]

I have argued for a very different interpretation of the oxygen controversies that dogged the British expeditions in the first half of the twentieth century: the debate was far less about ethics than it was about the difficulty of converting facts made in a laboratory setting to practical solutions on the mountainside.[38] Pugh toned down his own analysis less than a year later, in 1954, still blaming "the fact that [mountaineers] were extremely anxious to climb Everest by their own unaided efforts," but also acknowledging that the oxygen equipment that was used prior to the 1950s was not satisfactory.[39] There was genuine medical and scientific uncertainty about the use of oxygen on the mountain, and the opposition to an obsession with oxygen was not voiced uniquely by nonscientist mountaineers: as one example, we can turn to Dr. Longstaff, who, having written his article about mountain sickness being caused by fatigue, went on to be a vocal critic of the (heavy and therefore fatiguing) oxygen systems carried in the 1920s and '30s, particularly as he came to believe that a technical failure at high altitude could have serious consequences for an unacclimatized climber.

One serious setback for the cause of oxygen supplementation was the death of its key promoter on the 1921 British Mount Everest reconnaissance expedition. Alexander Kellas was a physiologist, doctor, one of the most experienced Himalayan mountaineers in early twentieth-century Britain, and the first European to recognize the superior climbing abilities of local people (he had taken two Swiss guides on an expedition to Kashmir and Sikkim in 1907, and both had suffered from mountain sickness).[40] Dr. Kellas had spent nearly fifteen years working on the physiology of altitude and respiration, and had collaborated with Haldane on low-pressure experiments, before being instrumental in the design of the equipment that was taken on the 1921 reconnaissance expedition.[41] Along with several other members of the team, he contracted what was probably dysentery on the trek in and died just before they reached Kampa Dzong in Tibet. The other expedition members survived, but Kellas's age (he was fifty-three), and the fact that he had just come from an exhausting research trip to Kamet (in Garhwal, India), may have led to his death. Kellas had been in Kamet on a trip sponsored by the Medical Research Council (MRC), the British Mount Everest expeditions' Oxygen Research Committee, and the (governmental) Department of Scientific and Industrial Research, studying the feasibility and function of oxygen technology at high altitude.[42] His death emphasizes that in a pre-antibiotic era, climbers were as much at risk from infection, nonpotable water, and gangrene as they were from avalanches and crevasses (another risk that, safe in the Sorbonne, Bert did not have to consider).

Despite this setback, oxygen was tried on all three of the British Mount Everest expeditions in the 1920s, and George Mallory and Andrew "Sandy" Irvine—both the epitome of the gentleman amateur—climbed to their deaths in 1924 wearing oxygen kits that Irvine had himself adjusted (see fig. 6):

> None of the oxygen apparatus would have worked if it were not for him, all the tubes being of porous brass which he has rendered non-porous with solder, etc. . . . If ever a primus-stove goes wrong, it goes straight to Irvine, whose tent is like a tinker's shop.[43]

The designs of the laboratory, however carefully tested in simulated conditions, do not always function as expected on the mountain.[44] The equipment was rarely satisfactory: in hindsight, it is clear that the flow rates were often too low to compensate for the extra weight and bulk of the equipment, and in addition, there were constant problems with fit, comfort, and ease of use, while important parts tended to freeze up, indicators failed, and masks fogged.[45] The expeditions of the 1930s had, if anything, less luck with their oxygen systems than those of the 1920s, yet despite these difficulties, the leader of the 1933 expedition explained that "the oxygen question was still open, and we could not afford to dispense with anything which might contribute to success."[46] Consequently, they still took oxygen systems, and they still spent time investigating and attempting to improve them. The teams of the 1920s and 1930s also made genuine efforts to do physiological work and to take clinical samples as high on the mountain as possible in order to provide data for the physiologists to analyze.[47]

But despite all this work, the fact was that by the outbreak of the Second World War, when expeditions ceased, the evidence in the field seemed to show that oxygen systems did not aid climbing—if anything, the additional weight, the bulk of the equipment, and the awkwardness of breathing through masks and nozzles seemed to make climbing at altitude more, not less, arduous. While the calculations of physiologists, even including oxygen proponents such as Kellas, suggested that reaching the summit of Everest must be very near the limit of feasibility, the hope remained that "men of exceptional powers of endurance" who were highly acclimatized could reach it without supplemental oxygen.[48] It was not until the early 1950s that a convincingly successful oxygen system was designed for British high-altitude mountaineering, as the result of research conducted by physiologist L. G. C. E. Pugh at high altitude and work done by the expedition's Oxygen Sub-Committee, both building on technological breakthroughs made during the war by scientists working on respiration systems for aerial warfare.

Some of the credit for the success of the teams of the 1950s can be given to increased funding for aviation-related research, which became such a focus during the Second World War.[49] (Pugh gained his first experience of altitude physiology in a military context when he became medical officer for the Mountain Warfare Ski Unit, which was created in the Lebanese mountains in 1941.)[50] But the relationship between military research and mountain climbing was not always helpful; the airplane cockpit may actually resemble the barometric chamber more closely than either effectively models the mountain. Those working on the British attempts on Everest were aware of this disjuncture, and it led to conflict, although this conflict has often been misunderstood as part of a battle between sport and science rather than what it really was: a scientific dispute about models and laboratories. This misunderstanding is epitomized by the phrase used as the subheading to this section, a quote that is most often (mis)used to illustrate claims about gentlemen climbers rejecting technology.[51] It is a paraphrasing of a letter written by the Cambridge astronomer Arthur Hinks, who represented the Royal Geographical Society on the Himalayan Committee, the group that had organized and sourced funding for all the previous British attempts on Everest. Hinks was writing to J. P. Farrar, the president of the Alpine Club, and the word "rotter" comes from this exchange:

I should be especially sorry if the oxygen outfit prevents them going as high as possible without it. The instructions laid down by Dreyer say . . . that oxygen should be used continuously above [7,000 m]. That . . . is all nonsense. . . . If some of the party do not go to [7,600 m] without oxygen they will be rotters.[52]

What makes this letter about something other than "amateurism" or the "ethics" of using oxygen is the name of Georges Dreyer.[53] Dreyer was a Danish medic and the first chair of pathology at Oxford University, collaborating with Barcroft and FitzGerald before joining the British Royal Air Force (RAF).[54] He was key in the development of the oxygen equipment used by pilots in the RAF—known as the Dreyer apparatus, it was also taken up by the US Air Service.[55] Dreyer's reliance on barometric chamber research and his specialist interest in aviation meant that his conclusions were not always shared by other physiologists or by mountaineers. Dreyer's suggestion that oxygen should be used continuously at the relatively low altitude of 7,000 m seemed to be contraindicated by physical experience on the mountain and by the research work on human acclimatization. Of course, acclimatization is not an issue for fighter pilots, who ascend far too fast for their *milieu intérieur* to adapt to a new environment; mountaineers, on the other hand, are normally able to provide some buffer against the lower P_{O_2} by autonomous physiological adaptations.

There was, as the mountaineers pointed out forcefully, another crucial difference between the pilot and the climber: the pilot's oxygen system was much more reliable than anything that had so far been taken up a mountain. Although early pilots were exposed to vast temperature changes, they were not independently mobile—the weight of their equipment was limited only by the power of their planes, not by their physical strength, and that meant that even heavy equipment could still benefit the pilot, while it would actively hamper the mountaineer. Likewise, the pilot was far less likely to bang, drop, or mishandle his tanks and valves, and they were at no risk of being caught on outcrops, cracked on overhanging rocks, or tossed into a valley by a recalcitrant yak. For a mountaineer, the use of oxygen at 7,000 m meant he was denying his body the opportunity to acclimatize, and therefore, if the set failed (and failures were frequent) at 8,000 m, his life could be at risk.[56] Far better to struggle on as high as possible, allowing the body's homeostatic systems to rearrange and rebalance, and to take oxygen only when it was essential.[57] In essence, the debate about oxygen was a debate about *necessity*, and about the reliability of facts made by people who had not climbed a mountain; it was not, in any meaningful way, dominated by the rhetoric of amateurism or the power of the gentleman climber.

War and aviation were not the only sources of knowledge on which the expeditions of the 1950s drew. Researchers also headed for South America to study *soroche* (the local name for mountain sickness). In 1921 Joseph Barcroft headed to Cerro de Pasco in Peru, a center of commercial mining over 4,300 m above sea level.[58] This town was thought to be the highest permanent human settlement on earth, while miners worked for months at a time at higher camps, at nearly 4,880 m. One purpose of this expedition was to deal with a controversy about the function of the lung: Did oxygen pass through the lung into the blood (and vice versa) purely through passive diffusion, or did the membrane of the lung actively "push" oxygen in one direction or the other? Haldane maintained that it did, Barcroft that it did not.[59] The Cerro de Pasco expedition would test this theory again, but "this time on more subjects and at *real* altitude in the field" (my emphasis).[60] It was also, however, designed to study long-term acclimatization and adaptation in a fairly rare population of high-altitude residents.

One unintended consequence of the expedition was that it acted as a direct stimulus, if not an active provocation, to local researchers—in particular, to Carlos Monge Medrano, who would later become the director (1934–56) of the Instituto de Biología y Patología Andina (est. 1931).[61] Monge M. had an active interest in altitude, and although he did not seem to know about the expedition in 1921, he did read Barcroft's 1925 book, *The Respiratory Function of the Blood, Part I: Lessons from High Altitudes*, and took great exception to one section, in which Barcroft says that "all dwellers at high altitudes are persons of impaired physical and mental powers."[62] Monge M. dismissed what he saw as a slur on the Peruvian people, arguing that Barcroft himself was suffering from mental impairment due to altitude and that his generalizations were unfounded. (Chapter 5 addresses the racial and national aspects of altitude research in more detail.) To disprove Barcroft, Monge M. organized his own expedition to Cerro de Pasco to study the exercise capacity of high-altitude residents, but in doing so, he went on to describe a pathological condition of high-altitude residents: chronic altitude sickness, subsequently referred to as Monge's disease.[63] He and his collaborators also created another node in a network of high-altitude research stations, allowing and encouraging international travel and collaboration.

Elsewhere in South America, Chile also acted as a host for expeditionary experiments, in this case conducted in the 1930s by members of the newly founded Harvard Fatigue Laboratory. Nominally intended to support research into fatigue as it related to industry (and, importantly, to relations between factory owners and factory workers), the Fatigue Laboratory interpreted its brief widely, and its researchers considered many ways in which environments could limit the work of the human frame, from dehydration to high temperatures to low oxygen concentrations.

In writing up the findings of the expedition he organized to Chile in 1935, professor of physiology Ancel Keys made the justification for this expensive overseas research trip clear: "The abnormality of life in [barometric] chambers makes them of limited utility except for acute experiments."[64] In any case, he went on, although the relationship between altitude sickness and reduced P_{O_2} seemed certain, there were still crucial research questions that remained unanswered in the 1930s, puzzles such as "the restoration of physical and mental vigour after some time at high altitude and the delayed onset of altitude sickness [which] are both more interesting and less intelligible."[65] The fact that Keys felt the need to explicitly explain the value of field over barometric chamber experiments should not be read as evidence of particular uncertainty about their worth; after all, many scientific papers begin with a justification for the choice of research method. But expeditions are certainly expensive (and sometimes risky) forms of science and perhaps require some justification, not least because expeditionary physiology is often supported by a mixed-funding model, so many organizations and institutions have to be convinced of its value before paying out.[66]

As a representative example, Keys's International High-Altitude Expedition to Chile—officially led by Harvard Fatigue Laboratory director D. B. Dill—was financially backed not only by the Harvard Fatigue Laboratory, but also the Milton Fund at Duke University, Copenhagen University, King's College Cambridge, Columbia University, the Royal Society (London), the Corn Industries Research Foundation, the Rask-Ørsted Fund, the Josiah Macy Foundation, and the American Association for the Advancement of Science. Nonfinancial support came from the Chile Exploration Company, the Ponderosa Mining Company, the Ferro-Carril de Antofagasta a Bolivia (a local train company), and many individuals.[67] While it is not unusual for scientific projects to have multiple sources of funding, in the case of extreme physiology these sources may include nonscientific organizations and atypical sources of science funding—including newspapers, climbing societies, the military, and pharmaceutical and other manufacturing companies as well as government funding. This money comes with terms and conditions that can limit or influence the expedition—in one case the funding was based on a promise to hunt for the yeti.[68]

While the scientific justification for trips to South American peaks often related to the need to study indigenous populations, or at least those who lived and worked for extended periods at mid- to high altitude, the Sherpa are conspicuously absent as objects of scientific scrutiny in the first half of the twentieth century. Kellas discussed the Sherpa's climbing abilities and recommended them as efficient porters at high altitude as early as the 1910s, and subsequent expedition reports do acknowledge the superior working ability of high-altitude Sherpa (and sometimes porters) compared with recently arrived European climbers. Yet no particular effort was made to study their physiology or cultural adaptation to heavy work at high altitude. It was not until 1952, when Pugh examined Sherpa during an Everest reconnaissance expedition on Cho Oyu, that Western science began to use local bodies for Himalayan research. Pugh's studies reported a lower-than-expected level of hemoglobin among the porters and high-altitude Sherpa, a finding at odds with what one might have assumed from the studies of high-altitude residents in the Andes (although in line with some of the disputed findings of Mosso and Jourdanet), as discussed further in chapter 5.[69]

From Everest to Silver Hut

The 1952 Cho Oyu expedition was the second specifically scientific expedition to the Himalaya funded by the British Himalayan Committee—Kellas's was the first, while the expeditions of the 1920s and '30s tended to test equipment in the Alps or during reconnaissance trips. Re-formed after the Second World War, the committee had begun to consider further British attempts on Everest, but without much initial sense of urgency. In 1951 the surgeon Michael Ward proposed a new route to the summit from the south (through the Khumbu Icefall and up the Western Cwm), and later that year a small reconnaissance expedition was sent to check the route's feasibility. Then, to the horror of the Himalayan Committee, the government of Nepal granted a Swiss climbing team, not the British one, access to Everest in 1952—and not just for one attempt on the summit, but for two, one pre- and one post-monsoon. The exclusive British access to the mountain no longer existed, and it became very clear that other nations would be lining up to take advantage of this opportunity to climb the highest mountain on earth. Therefore, the forestalled British plan to attempt Everest in 1952 was reconfigured into another reconnaissance, this time heading to nearby Cho Oyu, with the express purpose of testing oxygen equipment, conducting physiological studies of climbing, and improving other technologies, from crampons to rations. This intense research program was a response to fears that the next British attempt on the mountain might be the last.

There was another physiologist in the field in the 1950s: American doctor and experienced Himalayan climber Charles (Charlie) Snead Houston.[70] Houston's father, Oscar Houston, had obtained permission to enter Nepal in

1950, and the family—including Nelly Houston, Charlie's mother and Oscar's wife—got together in Kathmandu, where they met—by accident—the British climber Harold (Bill) Tilman, a veteran of the British attempts on Everest in the 1930s. (Charlie Houston had been part of Tilman's successful attempt on Nanda Devi in 1936, which set an altitude record.)[71] Supported by Sherpa, Tilman and Charlie Houston scouted the southern route up Everest, although they concluded that the Western Cwm was inaccessible—a conclusion later disproved by the 1953 British Mount Everest expedition. Prior to this visit to the Himalaya, Houston had been responsible for a major study on the problem of acclimatization, but its results proved to be unhelpful to those dealing with altitude sickness in the field. In 1944, using money from a US Navy program, Houston designed and ran an experimental protocol he called Operation Everest, in which four naval personnel spent thirty-five days inside a barometric chamber, in order to study the effects of lowered P_{O_2}, with and without supplemental oxygen.[72] Drawing on his own mountaineering experience, Houston aimed to mimic that experience as closely as possible by creating an "ascent profile" in which the pressure changes in the chamber were modeled on those experienced by climbers, including a gradual increase of altitude on the "trek in" and a "dash for the top." Operation Everest's findings were not promising for the British teams on the mountainside: at the pressure estimated to be equivalent to 8,848 m (the height of Everest), only two of the human subjects could stay conscious without oxygen, and only when they remained at rest. Physical activity, an actual climb to the summit, seemed an impossibility.

After the war Operation Everest had an influence on British planning when the Himalayan Committee re-formed. Wing Commander H. L. Roxburgh, of the RAF Institute of Aviation Medicine, wrote in a 1947 *Geographical Journal* article that "to climb Everest without oxygen . . . would be like climbing an equally difficult and dangerous but much lower slope in a state of alcoholic intoxication" (this use of *intoxication* was directly drawn from aviation studies in which the mental confusion of pilots had been recognized as a crucial factor in plane crashes).[73] Although an oxygenless ascent of the highest mountain in the world was not impossible in Roxburgh's view, it would require a man to be "physically and mentally exceptional and also remarkably lucky."[74] Roxburgh's article interested the Himalayan Committee enough for its members to invite him onto its Oxygen Sub-Committee. Here he joined L. G. C. E. Pugh; two Cambridge researchers, Prof. Sir Bryan Matthews and Dr. Winfield; engineer and veteran of the 1938 expedition Peter Lloyd; and physiologist Tom Bourdillon (who worked sometimes with his father, Dr. R. B. Bourdillon).[75] This committee was responsible for designing and testing the oxygen systems for the 1953 expedition.

Despite Roxburgh's presence, the Oxygen Sub-Committee was clear on the difference between mountaineering expeditions and aviation. While they acknowledged that "only the [Air Force] Institute of Aviation medecine [*sic*] has the resources to carry out" the experiments and tests necessary to develop functional oxygen systems, these systems had to be specifically designed for the needs of mountaineers and mountains, not aviators and airplanes.[76] Barometric chamber experiments could be controversial sources of evidence for those working on the problems of high-altitude mountaineering. First, there were obvious differences between the physical experiences of a chamber and a mountain: conditions in Operation Everest were relatively harsh, with four men sharing a chamber with a floor space only 10 feet by 12 feet, and 7 feet high, and crowded with beds and research equipment (added to which, two of the subjects smoked throughout the study).[77] There was little room for physical exercise, no change in temperature, no wind resistance, and so on. Perhaps even more fundamental was the issue of Houston's mapping from air pressure in the chamber to elevation on a mountain. Barometric pressure and elevation are, after all, two different measures of two different natural properties. In the first years of the twentieth century, Nathan Zuntz had discussed the problem of correlating barometric readings with altitudes, and in 1906 he and his research team published work containing what became known as the Zuntz equation for performing this conversion.[78] It was Mabel FitzGerald who tested the equation against data she had gathered in the field on Pikes Peak; her 1913 paper showed that it seemed to work in the real world, at least for the analysis of alveolar air samples.[79]

Despite Houston's experience on mountains, he chose to use models from aviation, rather than physiology, in order to transform air pressure into altitude. Houston used the International Civil Aviation Organization (ICAO) Standard Atmosphere, an estimation of the relationship between altitude and air pressure designed in 1924 for the calibration of aircraft altimeters. This choice was a puzzling one because as early as 1935, physiologists, first of all Haldane and J. G. Priestley, had pointed out that the Zuntz equation and the internationally agreed altimeter calibration standards began to disagree above 15,000 feet (4,572 m).[80] There are two main reasons for the disagreement between these figures, both having to do with averages and means. The ICAO Standard Atmosphere is calculated using average atmospheric conditions and a mean value for the depth of the atmosphere; in reality, the earth's atmosphere is unevenly spread around the globe and is thicker at the equator, thinning out toward the two poles. Because of their location near the equator, the Himalaya therefore have higher barometric pressures (because the atmosphere above them is thicker than average), and therefore higher P_{O_2}, than mountains of equivalent height located elsewhere on the earth—so if Everest were located in Western Europe, it might not be possible for humans to climb it without oxygen. Temperatures also make a difference to baromet-ric pressures, and in the warm seasons, when mountaineers tend to attempt high-altitude climbs around Everest, the discrepancy can be considerable: at the summit of Everest, the assumed "standard temperature" overnight for that altitude would be –40°C, while the observed temperature in 1953 was, on average, –27°C.[81]

Houston's calculations were for a colder mountain, located closer to West-ern Europe than to the equator; the four men in his chamber were experienc-ing the atmospheric pressures one would expect to find not on the summit of Everest itself, but perhaps as much as 180 m above it.[82] Any attempt to map such abstract measurements to real-word heights was going to be problematic in 1944, as the first direct measurement of the air pressure on the summit of Everest was not taken until 1981. As late as 1970 D. B. Dill and his co-researcher D. S. Evans were still pleading with the readership of the *Journal of Applied Physiology* to "Report Barometric Pressure!" more consistently in papers about high-altitude expeditions so that better calculations could be made.[83] Despite these pleas, by the end of the century only two direct mea-surements had been made on the summit of Everest.[84] The scanty readings that were taken served only to highlight another problem with barometric chamber models: pressures on real mountains are incredibly variable.[85] Read-ings on Everest range from around 243 to 255 torr, depending on the weather; to give a sense of scale, that equates to a difference in altitude of 300 m—so in good weather Everest is "only" the equivalent of 8,700 m above (average) sea level, and on a bad day it may be as high as 9,000 m.

The variation in barometric pressure readings on real mountainsides was what gave hope to a British scheme to climb Everest without oxygen in the 1960s. In a "plan for a combined mountaineering and scientific expedition to Everest," written around 1959, Pugh spun this variability as a positive. He pointed out that since the 1930s many climbers had successfully reached 28,000 feet (8,534 m) above sea level without oxygen, and that

> the difference in barometric pressure between 28,000 and 29,000 feet is a mere 10 mm Hg, and the difference in the partial pressure of O_2 in the in-spired air would be only 10 × .21 = 2.1 mm Hg. Meteorological fluctuations in atmospheric pressure must be at least as great as this at these altitudes.[86]

The empirical experience of mountaineers, and the uncertainties of the natural world, contradicted the findings of the barometric chamber—a gasless ascent was possible. (Pugh's analysis was not always consistently optimistic; just a year earlier he had published a paper saying that it was "unlikely that the mountain could be climbed without oxygen equipment without serious risk.")[87]

Pugh's case rested on the need for a faster ascent to the summit than previous teams had managed, a real dash for the top. This, he suggested, might be possible only if climbers were thoroughly acclimatized to altitude, or at least had achieved the highest level of acclimatization possible for a person born at sea level. That was the principle behind a scientific expedition co-led by Pugh and (Sir) Edmund Hillary in 1960–61. With "political difficulties" standing in the way of an attempt on Everest, the world's fourth highest mountain, Makalu, was chosen as an alternative goal. An extensive scientific program was designed for the expedition, whose official name was the Himalayan Scientific and Mountaineering Expedition, 1960–1961, but which has subsequently become known as the Silver Hut expedition because of the iconic silver laboratory used as a base for the physiological studies.[88]

The Silver Hut expedition is an example of the interactions between Antarctic and high-altitude research, as the concept of the expedition was based on the Antarctic practice of "overwintering"—that is, the use of extended stays in the Far South to allow a variety of exploratory and scientific goals to be achieved. Its borrowing from, or building on, previous expeditions is acknowledged in the grant proposals; in one to the Wellcome Trust, Hillary makes reference to the work of the International High-Altitude Expedition to Chile, pointing out that "twenty-five years have elapsed since the last full-scale physiological expedition to high altitude."[89] While this claim somewhat glosses over Pugh's work in the 1950s (and the relationship between the two men was to develop into open hostility and conflict during the Silver Hut expedition),[90] Pugh himself made connections between work in South America and in the Himalaya. He wrote enthusiastically about the possibility of studying Sherpa and comparing their physiology with that of the indigenous populations of the Andes—and indeed, conducted some of the first tests of Sherpa physiology, which, as chapter 5 will show in more detail, overturned some of the assumptions physiologists had drawn about adaptation to altitude.[91]

Initially, the Silver Hut expedition had two key aims. The first was to carry out a comprehensive program of scientific investigation using mountaineers' and scientists' bodies, facilitated by a long stay at altitude and the construction of semipermanent laboratory and accommodation huts. Second, the long stay at altitude and (hoped-for) accompanying acclimatization was supposed to allow for an oxygenless ascent of Mount Makalu, which, at 8,485 m, would have been the highest peak scaled without supplemental oxygen. To this plan, an additional activity was added to keep some of the climbers occupied during the acclimatization period and to raise crucial additional funds: a hunt for the yeti. This endeavor was a completely serious one, planned in extraordinary detail by Hillary with help from Marlin Perkins, the director of Chicago's Lincoln Park Zoo, who handled some of the more exotic technology, such as long-range dart guns and a tear-gas pencil. Not all the climbers were entirely convinced about this technology—New Zealand climber Peter Mulgrew, perhaps in an echo of the debates over the *practicality* rather than the *ethics* of oxygen use, later wrote rather sarcastically that

> all one had to do on coming face to face with a Yeti was to estimate the body weight of the creature, carry out a small mental calculation, enabling one to set the correct amount of drug so as not to kill the beast, load the gun and fire. . . . In any event I found myself quite unable to do the mental sum required and on the one occasion when I did manage it, took a full six minutes.[92]

The three aims of the expedition led to complicated logistical arrangements and a fairly large team of yeti hunters, climbers, and climbing scientists. In the latter group were six physiologists who overwintered in the hut for at least part of the season—Pugh and two other British physiologists, James Milledge and John West; the New Zealand climber and medical student Michael B. Gill; US Air Force doctor Tom Nevison; and Sukhamay Lahiri, an Oxford University–trained Indian physiologist—plus a geographer (Barry Bishop), a captain in the Indian Army Medical Corps (S. B. Motwani), and a skilled carpenter (W. Romanes).[93] While Hillary led the yeti hunt, Milledge, Bishop, Romanes, mountaineer Norman Hardie, and 310 porters moved the scientific equipment to the Mingbo Glacier and erected the huts: Green Hut for living and Silver Hut for experiments.[94] The overwintering team arrived in late 1960, and from this team, Gill, Romanes, West, and Milledge joined Hillary's climbing team on Makalu for a disastrous summit attempt, during which Hillary suffered a stroke and a series of accidents resulted in the eventual amputation of both of Mulgrew's feet.[95]

Despite all the possible conflicts of interest, and the quite real conflicts between Pugh (and others) and Hillary over the latter's organizational skills and "reckless" approach, the scientific program was extraordinarily ambitious and successful.[96] The effects (or lack thereof) of acclimatization on the whole body—from body fat to respiratory capacity, from adrenal secretions to blood volume—were tested over a long period, and active experiments using bicycle ergometers, step tests, psychological card-sorting tests, and electrocardiograms were used to study the long- and short-term effects of altitude on the human (male) body.[97] Later physiologists have referred to Silver Hut not only as a seminal moment in extreme, altitude, and respiration physiology, but also as a set of experiments that produced results of which, over fifty years later, "none ha[d] been refuted"; several of the measurements remain the "highest ever" even to the date of this book.[98]

Children of Silver Hut

The Silver Hut expedition was extraordinarily productive by any measure, the most obvious of which was the publication of at least forty papers.[99] Among this mass of data, three results are particularly important for the stories being told in this book. The first was the finding that there was no change in the structure of the climbers' lungs when it came to their diffusion properties (i.e., the rate at which gases can pass from the alveolar space to the blood, and vice versa). This finding was in stark contrast to work on populations resident at, or indigenous to, high altitude, who do show diffusion rates that differ from the norms of sea-level populations. Its implication was that some adaptation to altitude is inherent, genetic, racial, and perhaps unachievable for climbers with sea-level ancestry. This conclusion was emphasized by a surprise second finding: long-term acclimatization did not give climbers any noticeable physical advantage over those who had experienced a shorter period of acclimatization during the weeks of a trek into Base Camp. If anything, extended stays at mid- to high altitude might actually have decreased performance, as shown by a surprising third finding of the Silver Hut experiments, where for the first time, a decline in maximal exercise ventilation was measured among team members. (Maximal exercise ventilation is the maximum volume of gas a subject can inhale and exhale during exercise and is therefore a limiting factor in the amount of physical work he or she can do.)

The significance of this last result is not just its general physiological implications, but rather that it was replicated in later field studies at high altitude, but *not* replicated in later barometric chamber studies; here was yet another difference between the real world and the laboratory, another way in which the model of the mountain failed to produce true facts about the actual experience of mountaineering. It was also, at best, an ambivalent result when it came to the question of finding the limits of human survival at altitude: the disaster on Makalu prevented a world record for an oxygenless ascent, and the fact that long-term residency did not improve acclimatization must have cast some doubt over the idea that any climber could become well adapted enough to make a suitably fast summit climb (in fact, as some subjects seemed to suffer continual weight loss despite adequate diet, long-term residency at altitude might actually be harmful in several ways). But hope was not lost; after all, one of the repeated lessons of such fieldwork was that models do not always accurately represent reality, and Silver Hut was, after all, a model. Although at high altitude (5,800 m), it was not located as high as Everest: facts about heights above 7,440 m, where the highest measurement was taken on this expedition, could only be extrapolated, not demonstrated.[100] Hillary himself believed that an oxygen-less ascent of Everest was still possible, and he claimed in 1976 that he had "always" believed so.[101]

Only Everest itself could turn this theory (or this belief) into a fact—and that happened in May 1978, when Peter Habeler and Reinhold Messner became the first men to reach the summit of the mountain without supplemental oxygen. Of course, what their success showed was that man could climb the mountain; it did not indicate whether his ability to do so demonstrated a failure of the ICAO (and similar) models to map altitude onto barometric pressure or a failure of the physiological theories used to understand human survival at altitude. When the first barometric reading was actually taken on the summit of Everest three years later, "the physiologist . . . [was] somewhat relieved to find that the barometric pressure is indeed elevated above the standard US Atmosphere value . . . for that reduces the problem of explaining adequate oxygen supply for survival."[102] In other words, a mistake had been made by physicists and mathematicians, not by biologists and physiologists.

The productivity of Silver Hut was not just in its papers and data; it also created careers and networks, as many of the men involved in the research went on to become key figures, internationally recognized as such, in altitude, respiratory, or exercise physiology. The ways in which unusual research sites, such as Everest, can create tightly woven networks of researchers is the topic of chapter 3, although it is worth emphasizing here that these are almost entirely male networks. While female climbers were present in the Himalaya by the mid-1950s, they were not usually part—either as climbing subjects or climbing scientists—of Western scientific expeditions until the very end of the twentieth century; a few notable exceptions are also discussed further in chapter 3. Other countries, particularly the communist states of China and the USSR, systematically had female scientists on mountains before the Western nations, although they were apparently doing meteorological and geographical research rather than studying their own (or others') bodies.[103]

Silver Hut also appears to have inspired other expeditions, and while the focus here is on the specifically physiological expeditions to the Himalaya, it is worth acknowledging that a fully international cast of scientists were making use of the men (and later women) who were willing to attempt to climb the world's highest mountains, and that some physiology work was undertaken as a part of more generally scientific expeditions (e.g., the American Mount Everest Expedition of 1963).[104] Many expeditions reported back on experiences with oxygen systems or altitude sickness, and in some cases made (generally quite basic) clinical and physiological observations that were later published; for example, a fairly comprehensive study was done on the climbers of the 1973 Italian team on Everest, led by the extremely wealthy newspaper owner Guido Monzino. While Monzino reputedly had a "carpeted five-roomed tent equipped with leather upholstered furniture" at Base Camp, a less glamourous, but nonetheless "comfortable," 4 m² tent was also provided for scientific work at the top of the Khumbu Glacier (5,350 m above sea level).[105] Here, body weight, arterial blood pressure, red blood count, hematocrit, hemoglobin concentration, arterial O_2 saturation and O_2 and CO_2 partial pressures in the blood, blood pH, blood lactate, ventilation rate, and heart rate were all measured and compared with sea-level control studies done at Milan.[106]

Many expeditions conducted such studies, although perhaps on smaller scales, and most either confirmed, or slightly extended, the studies in Silver Hut; the physiologist Paolo Cerretelli's work on the 1973 Italian expedition's data showed, for example, that even when well acclimatized and given pure oxygen, climbers often do not show the same level of physical efficiency that they demonstrated at sea level. The Himalaya were also a popular destination for the more ambitious university climbing and exploratory clubs. Jim Milledge's alma mater, Birmingham University Medical School, formed the Birmingham Medical Research Expeditionary Society in the late 1970s, which undertook its first expedition to western Nepal to investigate acute mountain sickness (AMS) in 1977. It was the first British expedition to study AMS, and the participants published a series of papers about the symptoms, experiences, and clinical data generated by seventeen subjects, all of whom were male.[107]

Despite the many expeditions that made gestures toward research by taking clinical measurements, by including doctors and physiologists, or by signing up members for clinical examinations before and after, it was not until 1981 that another temporary non-tent laboratory was used at high altitude in the Himalaya. The American Medical Research Expedition to Everest (AMREE) seems the most obvious sequel to Silver Hut, not least because it was organized by Silver Hut participant John West, along with Dr. Chris Chandler and Dr. F. Duane (the full roster of participants and advisors for the AMREE reads like a roll call of the key names in respiration physiology and mountaineering, including Nello Pace, Hillary, Milledge, Lahiri, Houston, Dill, and Thomas Hornbein).[108] Documents and grants relating to the AMREE certainly claim Silver Hut as an ancestor, although we should bear in mind that there is a strong nostalgic thread in such mountain work that is not always present in other areas of research (more on this in later chapters). Aside from the basic principle of doing physiology at high altitude in the Himalaya, it is the concept of the mobile laboratory that AMREE most obviously borrows from the Silver Hut expedition. It also, perhaps unconsciously, borrows its three-part structure, organizing the relatively large team into six dedicated climbers who were supposed to create a route to the summit, six "climbing scientists," and eight "scientists" who were to man the two laboratories but not go above 6,300 m. Three of the Americans successfully summited, and two of these were climbing scientists Dr. Chris Pizzo and Dr. Peter Hackett. Pizzo not only recorded the first direct barometric pressure reading on Everest, but also undertook the highest clinical investigation, taking samples of his own alveolar air while sitting on the summit.[109]

The tension between the laboratory and the field had not abated, even a century after the first critiques of Bert's work. Replying to a query from a colleague in the Physics Department at the University of California, San Diego, John West wrote (rather patiently) to explain why barometric chambers could not be used "instead" of the expensive mountaineering trip:

> The question that you pose . . . is very often asked. The answer is that in order for people to tolerate these extremely low levels of oxygen, they need to acclimatize at high altitudes for a period of approximately 2 months. . . . I looked into [long stays in chambers] with the people at the US Army facility in Natick, Mass. which is the most elaborate low pressure facility in the country, and they doubt that it is feasible. . . . Probably setting up an experiment in a low pressure chamber would be considerably more expensive than using the natural laboratory on the mountain.[110]

Further, the ethics of the experiment had to be considered: West went on to point out that the only time such extensive studies had been done in a chamber was for Operation Everest, and "[as t]his was done during wartime . . . the naval recruits had little option but to agree."[111]

Peacetime long-term barometric studies were possible, but extremely rare. At the same time the AMREE was taking shape, Charles Houston, the lead experimenter in Operation Everest, was designing a new barometric chamber experiment. More than forty years after his original study, Operation Everest II (OEII) tested eight men over a forty-day residence. It did not prove easy to fund this investigation, as Houston's initial grant bid to the National Institutes of Health was rejected, but the US Army Medical Research and Development Command eventually agreed to fund the operation. It was possible, in the chamber, to do invasive tests on the bodies of the eight men, including cardiac catheterization, and even the most ardent proponents of fieldwork in extreme physiology admitted that these sorts of investigations were simply not practicable at high altitude on mountains. (Houston had justified his first Operation Everest in part by claiming that "since oxygen lack notoriously dulls cerebral function," the observations of climbers were "open to question"—he did not repeat that suggestion in the 1980s.)[112]

The conversion between pressure and altitude remained controversial, especially as the experimental design of OEII introduced more sources of error; for example, when the experimental team entered the chamber to do the catheterizations, they were using supplemental oxygen, some of which escaped into the chamber itself, altering the P_{O_2} and effectively "lowering" the altitude. West wrote to Houston to criticize the use of estimates of altitude, rather than measured barometric pressures, in published papers.[113] These criticisms appear to have had an effect, as earlier OEII publications contain references to altitude, while later ones switch to relying on barometric pressure; those published in the late 1980s also explicitly point out the differences between barometric chambers and, as they term it, experiences on "real mountains."[114]

Operation Everest III took place in 1997, this time closer to Bert's original "climb" of Everest, as it was run by French researchers at the COMEX facilities in Marseilles, France.[115] By the end of the twentieth century, the barometric chamber was again regarded as a relatively good model for the mountain, although, ironically, it was no longer thought to be a good *alternative* to the mountain as a research space. Any claim that the barometric chamber, as a purified and reified laboratory practice, eliminated "complicating" factors of the mountain had been conclusively disproved. Subjects in chambers also experienced fatigue, distress, sore throats, disrupted sleep patterns, and so on, and subjects' physiological measurements (in terms of heart rates, respiration, and hormone levels) seemed very similar to those of mountaineers, meaning that the "'mountain stress factors' were not significant confounding factors for studies conducted in the field."[116] Consequently, as expeditions had actually become cheaper than barometric studies by the 1980s, if not earlier, a special case now had to be made for choosing the laboratory over the field, and not vice versa.

The Unpublished Truth about Fields and Laboratories

It would be too simple to characterize the divisions between Operation(s) Everest and AMREE as a conflict between two ways of doing science: laboratory and field. While many of the experimenters who used barometric chambers and exercise bicycles never got to work out on the mountainside, all of the field physiologists spent time in sea-level laboratories. Pugh is an excellent example, as any of the research questions that attracted his interest (which ranged widely outside of respiration physiology, as we will see in later chapters) could be addressed with benchside experiments; chemical analyses performed off-site by private or public laboratory facilities; collection of anecdotal and subjective descriptions of performance or symptoms by athletes and explorers; his own personal experience; field studies in semistructured sites (such as high-altitude laboratories or sports centers); and exploratory science, taking measurements wherever he could on a mountain or in Antarctica. In this sense, Pugh's work looks much more like the bricolage of Mosso's work than the single-minded, single-approach work of Paul Bert. This melding of methods is perfectly illustrated by Pugh's laboratory: in 1967 he successfully persuaded his employer, the MRC, to put him in charge of a laboratory of his own, which was called the Laboratory for *Field* Physiology. Movement between research spaces, from sea-level laboratory bench, to mid-altitude temporary laboratory, to frigid, windswept mountainside over 7,500 m above sea level, was not always simple and seamless; it was intellectually and practically challenging to relocate and synchronize the movement of researchers, research subjects, and pieces of material culture (and we will look at the challenges of mobile and circulating equipment and biological samples in chapter 3). There were clearly limitations on the work that could successfully be done in the field. The mountain and, as we shall see later, the Antarctic and other field sites may have been "natural laboratories," but they were also high, cold, dangerous, distant, and unpredictable.

However, what extreme physiology demonstrates is that the limitations on fieldwork are not necessarily the ones historians (or scientists) have expected. That is, the difference between the laboratory and the field is traditionally figured as a difference of control: the bench experiment is fundamentally about known quantities and conditions, about a controlled and manipulated environment, about a simplified and scaled-down model that can exclude confounding factors. Field sites are messier, producing less certain results, or results that cannot always function as generalizable facts about nature (so a field study might be thought to have produced facts about the particular, individual, field site studied, but not about other, similar, physical sites, ecosystem interactions, embodied experiences, etc.). Sometimes this was the case: oxygen-deprived mountaineers did not always report reliably about their experiences on the mountain, and even in the rigorously planned Silver Hut program, one experiment was seriously disrupted because "a Sherpa had been using a primus stove inside the Silver hut which was well sealed," causing the participants to suffer from carbon monoxide poisoning.[117] And, as we have seen, facts made on mountainsides did not necessarily apply to other sites, such as the cockpit of an airplane. But these limitations were in no sense the major features of the experimental work done in extreme environments, and, perhaps more importantly, they were also present in the laboratory sites. Experimental models, such as barometric chambers or running machines, had to be constantly tweaked and improved so that their results conformed to data gathered in the field. They were also just as vulnerable to confounding factors, to a loss of control—an example is the leaking of experimenters' masks in Operation Everest II (OEII). Even at the end of the twentieth century it was still profitable for physiologists to ask the question "Are laboratory and field [observations] related?"[118]

Many of the facts about human bodies produced by high-altitude expeditions were generalized to other situations, and as we will see later in this book, the results of experimentation in extreme environments were thought to be relevant to a host of other situations, from space flight to incubators for premature babies. Conversely, the facts made in barometric chambers could sometimes confuse issues if generalized to the mountain. For example, the researchers involved in OEII described their method as a "simulated climb of Mt. Everest."[119] That is, their literal claim was that their experiment simulated not a specific barometric pressure, or even a specific altitude, but a climb *of a specific mountain*. This claim is clearly unwarranted, not just because of the problem of pressure/height conversions, but also more pragmatically because (aside from a few step tests) no one climbed anything inside the chamber, much less had an experience that could meaningfully be described as "more like" an ascent of Everest than of, say, Cho Oyu or Makalu. It would be pushing the fidelity of the model to say that the participants "ascended to 8,848 m," let alone "climbed Mt. Everest."

Giving the researchers the benefit of the doubt suggests that the comparison is being used, much in the same way Paul Bert used it over a century before OEII, as a familiar comparator, a proxy for height that emphasizes the extremity of the research, rather than as a claim to *literal* experience. Nonetheless, this linguistic slippage between real-world measures, models, and approximations could clearly be problematic.[120] Of course, researchers "in the know" would not have been led astray by the OEII papers—they would have known, because it had been so widely discussed since 1935, that barometric chambers can replicate only pressures, not altitudes. Presumably, their objections were more focused on the possibility that this language could be misleading for people outside or new to the discipline, as well as reflecting their frustration when they had to convert altitude back into pressure in order to use or reanalyze the data from OEII.

Such "insider information" performs a vital role in expeditionary science. Significant pieces of information are left out of the official reports of expeditions, so while the publications from Silver Hut are more careful than those from OEII in their use of altitude and pressure, and while they are systematic and detailed, often describing errors or serendipitous findings, they still miss a huge crucial part of the researchers' methodology: how to get to the Mingbo Glacier in the first place. They do not detail how to organize and fund an expedition, what rations to take, how to choose good guides and Sherpa, how to negotiate import permissions for dangerous chemicals or oxygen cylinders, or what to wear in order to stay warm and yet maintain the mobility necessary to build huts, take blood pressures, or dismantle exercise cycles. These procedures are omitted despite the fact that reports of expeditionary science often do involve far more detail about environmental conditions, laboratory space, and survival mechanisms than one finds in most papers in physiology journals.[121]

A vast amount of the knowledge needed to run a successful field experiment at high altitude came not from published scientific papers, reports, or even edited volumes and conference papers, but from three other sources: a researcher's own previous personal experience, testimony from other people with personal experience, and nonscientific mountaineering literature, including popular books. This knowledge was still cumulative and adaptive: for example, the experiences of the British Everest expeditions of the 1920s and '30s, transmitted through reports, unpublished papers, and personal testimony, were crucial to the work of Pugh and the rest of the Oxygen Sub-Committee in the 1950s. Likewise, Pugh spent four days in Geneva debriefing with the Swiss team in 1952, sharing his Cho Oyu findings with them in return for "detailed inventories of their clothing, equipment, and diet, along with written accounts of their experiences."[122] In turn, when physiologist Tom Hornbein was tasked with organizing the oxygen technology for the AMEE in 1963, he wrote to, and got samples from, the Swiss team and from Dr. John E. Cotes, a researcher who had helped design the masks and tubing for the 1953 British Mount Everest expedition.[123]

This sharing of information was not entirely between private individuals or research teams. Many of the companies that donated materials to the

British Everest expeditions expected feedback on the performance of the products at altitude. Some of this was clearly for advertising purposes, but some fed back into product development.[124] The (Swiss-)American mountaineer Norman Dyhrenfurth claimed that shortly after he returned from the 1952 Swiss attempt on Everest, he was contacted by the American firm Union Carbide, who wanted "to find out more about the various types of oxygen equipment which had been used by European Himalayan expeditions."[125] Dyhrenfurth turned this approach around in 1961, writing to Union Carbide suggesting that they should get in touch with Hornbein to assist in, or learn from, the improvements he was trying to make to standard French and Swiss equipment for the upcoming American attempt on Everest. In turn, Hornbein appealed to commercial interests as well as national pride when he approached Sierra Engineering in 1961 in the hope of encouraging the firm to help him solve some of the problems with existing oxygen sets (mostly relating to the accumulation of ice and increased resistance in the valves when the wearers were doing heavy work and breathing hard).[126] Sierra Engineering did not respond, so the AMEE relied on European equipment, with Hornbein favoring a Swiss-designed system based on the British 1953 sets, but with a French "wire-wound-alloy bottle" system (indeed, European equipment dominated high-altitude climbing until the 1980s).[127] This sharing of information had actually led to an informal standardization of the oxygen equipment, somewhat to Hornbein's frustration, as it turned out that the masks he'd requested from three different expeditions were all essentially the same:

> The French mask came. It is not radically different than the original ine [sic]
> I have from Masherbrum, is little better in function and not a whit less of a
> problem from freezing. My hope was the mask might possess a different fit
> but it turned out to be the same RAF type mask as the Swiss used and also as
> the British modified for their sets.[128]

All of the expeditions discussed in this chapter included a long stage of planning in which letters were written to previous expedition leaders, members, and suppliers soliciting unpublished information. This information included feedback on the performance of technology, on the choice of other equipment and supplies, such as food, clothing, and tents, and general advice on preparation, planning, and pre-expedition testing for any new or adjusted items. Some information found its way to the teams by accident; for example, in 1961 Hornbein mentioned to Dyhrenfurth that he had bumped into "John West, a young physiologist passing through . . . on [the] way back to London from Makalu"—that is, from the Silver Hut expedition. West had told Hornbein about an airstrip that had been constructed "at about 13,000 feet within a days [*sic*] march of the Everest base camp," a fact that Hornbein thought "useful knowledge from a medical point of view if none other" (presumably for emergency evacuations or supplies).[129]

Knowledge about the reactions of the human body to high altitude, and expertise in studying these reactions, gradually became consolidated (largely through experience) in individual researchers, but it also became manifest in objects. As described above, the gas masks used by one team—sometimes those that had actually been taken up mountains, sometimes spares, sometimes duplicates—could be passed on to a new team, who would adjust, tweak, or reinvent the technology, which would in turn be passed on to future climbers and scientists. Similar things happened to research facilities: Silver Hut, in particular, which was donated to the Indian government, became a key site for the Indian Defense Institute of Physiology and Allied Sciences (DIPAS), which was founded in part to coordinate and make better use of the facilities in the hut.[130] As a "passage point" for Indian researchers interested in altitude physiology, Silver Hut's appeal is not just its facilities, but also the connection it provides to the "seminal," "historic" studies of the 1960s and to the men usually considered the founder figures of the field.[131] "Trekking to Silver Hut," wrote a leader of scientific trial teams for DIPAS, "is almost like a pilgrimage, paying homage to the stalwarts of the erstwhile Silver Hut Expedition."[132] For the fortieth anniversary of the Silver Hut expedition, Jim Milledge drew up a plan for the "the 'Silver Hut survivors' to trek in to the site, to review some of the results of the expedition, and to revive old memories"— but the team went not to the Mingbo Glacier, but to the new site of the (by then quite tatty) Silver Hut, in Chowri Kang, about 4,350 m above sea level (well below the 5,800 m of the original location).[133]

The idea that a field site can maintain its status as a place for producing knowledge, even when it has been removed from its original location, is a rather complicated proposition. After all, what made Silver Hut special in the 1960s was its location at high altitude in the Himalaya; when it came down to Chowri Kang, it was lower in altitude than the Capanna Margherita (4,550 m). Capanna Margherita had mostly been used for physics and meteorology through the mid-twentieth century—for example, for cosmic ray research—but in the late 1970s, a few years after Silver Hut was relocated to Chowri Kang, it was renovated and became the base for physiological studies once again. Twenty-first-century expeditions make use of it: the Caudwell Xtreme Everest Expeditions and other teams from the Centre for Altitude, Space and Extreme Environment Medicine, based at University College London, use the renovated hut as a staging point for teams intending to go to higher altitudes.[134] Researchers have to argue for their place on the competitive, expensive Himalayan expeditions, and they are weeded out through a test of the practicability of their work at sea level, and then another at the Capanna Margherita.[135] Consequently, the mid-altitude hut can function as a gatekeeper for those scientists who want to take their work higher. This gatekeeping work, the ways in which access was allowed and denied to certain people, how networks of people and objects were created, and how extreme environments contribute to creating experts and expertise, are the topics of the next chapter.

THREE

Frozen Fields

The scientists who work in extreme environments are often explicit about the physical and mental challenges of their work. When Dr. Chris Pizzo took his record-breaking alveolar air samples on the summit of Everest in 1981, he described the stiffness of his equipment and the dizziness he felt exhaling into the air sampler.[1] In fact, he nearly did not make it to the summit at all; violent storms trapped the AMREE summit party at Camp V, and in the course of the storm Pizzo lost his ice axe. This loss was a serious setback to his hopes of making the summit—the ice axe is a crucial piece of equipment for the last push from the South Col—but he continued to climb, armed with a tent pole instead. Barely a hundred feet from the camp, he discovered an ice axe lying in plain sight; he picked it up, made the summit, and took the first ever physiological measurements at 8,848 m above sea level.

John West later described the finding of the ice axe as a "million-to-one chance" that was "largely responsible for the expedition obtaining the extremely valuable scientific information" from the summit.[2] As Pizzo later found out, he had a woman to thank for his "miraculous" find: two years previously, in early October 1979, West German mountaineer Hannelore Schmatz's oxygen supply ran out, and she died just above the South Col. Her body, frozen in an upright position, remained on view on the mountainside for years (in 1984, two men died trying to remove her corpse during a cleanup operation, and in the end, natural forces tipped her body into a crevasse).[3] It was Schmatz's ice axe, abandoned in the snow, that appeared at just the right moment for Pizzo. Even more dramatic encounters with the past—even with the dead—happen frequently in extreme environments; on the very same AMREE expedition Dr. Peter Hackett, having summited alone, had a serious fall, sliding uncontrollably down the Western Cwm (an unquestionably fatal plunge of around 2,800 m) before catching his boot in an outcrop. Trapped upside down, with the light fading and no support due until the next day, he "miraculously" found a rope fixed by a previous climber and pulled himself to safety.

Although the slopes of Everest, like the polar deserts, are romantically represented as natural sites resistant to human occupation, the reality of exploration and expedition is that in many cases multiple explorers are traveling the same routes—often the only safe or practical routes—through otherwise unexplored wildness. By 1911, in the case of the Antarctic, and 1922, in the case of Everest, traces of human survival and scientific work were repeatedly encountered. In some cases these traces were deliberate caches or depots, items left for one's own team members or for imagined future explorers, and were part of creating a trusted international community. In others, as with Schmatz's ice axe or the fixed rope Hackett found, they were items accidentally lost or abandoned—providing crucial resources in places with scarce natural supplies and occasionally saving lives "miraculously." Such traces provide links between expedition teams, and between individuals, who have never met (who, indeed, may have been born decades apart). Recovering and restoring abandoned technology and material culture can also function as a way to claim a link to a heroic or mythologized past.

Successful expeditionary science relied heavily on such informal and directly personal methods of communication. As the previous chapter showed, it was commonplace for teams planning expeditions to correspond widely with previous explorers—asking not only for their written expertise but also for material objects, such as gas masks or ration samples. Both experimental and experiential knowledge—categories that frequently overlap in extreme environments—were passed through a variety of channels. Ideally for the historian, these ideas would be shared via letters that ended up in an archive somewhere, but perhaps more often they would be passed on through contacts that are much harder to trace—incidental conversations over a cup of coffee at a conference or a hot buttered tea in a Tibetan monastery. These personal connections were a two-way street: as Hornbein sat in Washington University, St. Louis, desperately trying to improve the gas masks for AMEE by writing to scientists in Europe, he was not just gathering information from them, but also opening up a route for them into the American community of researchers and explorers. Similarly, because of the way expeditions were organized, an enthusiastic novice climber or recent medical school graduate could, in five or ten years, be an expedition leader or major grant holder;[4] thus helpful advice given to young researchers early could provide rewards later on. Just a few years later, Hornbein's advice on respiration systems was being sought out by German, Japanese, and Indian expeditions to the high

Himalaya. Arctic historians have clearly traced the "genealogy" of British Arctic expeditions in which the youth of one party became the leaders of the next, passing on their experience.[5]

Young researchers clearly recognized expeditions as opportunities. In 1958 one recent medical school graduate who had secured a place on a British-Italian Himalayan expedition (as team doctor) wrote to the British physiologist L. G. C. E. Pugh to ask for advice on designing a research project while he was in the region. With an eye to a research, rather than a clinical, career, he asked, "Do you think there are any medical observations I could make to further low temperature high-altitude research and possibly be of value to me in obtaining a suitable post later on?"[6] Such attempts to break into the field of extreme physiology emphasize the fact that it was often an exclusive form of scientific practice, with restrictions on participation that were economic, social, cultural, and political. This chapter looks at the ways in which this scientific "field" was frozen—that is, how certain people were allowed access to the sites of extreme physiology research, how expeditionary teams were created and maintained, how claims to authority were reinforced, and how a global network of people, ideas, and material culture was created to support the work of expeditionary scientists.

Exclusive Experiences

The extraordinary demands—in terms of time, money, and physical ability—of expeditions to extreme environments meant that they were exclusive spaces until well into the middle third of the twentieth century. The essential nature of informal communications—the need for personal contact from expedition to expedition, even if just via a letter—assisted in the creation of a clique of scientist-explorers with a rare combination of experimental and experiential knowledge.[7] The latter was particularly crucial: choosing an unreliable colleague could result in costly errors, cause the failure of experiments or whole expeditions, or, at worst, be deadly. In the records of expeditions, the desire to choose men who were the "right sort" is restated again and again. The subjective component of these choices has been picked over by social historians, who find in them explicit class and national prejudices (and, of course, clear gender discrimination). Biographers and historians have criticized Robert Falcon Scott for allowing Apsley Cherry-Garrard to buy a place on his doomed *Terra Nova* expedition to Antarctica, given Cherry-Garrard's inexperience and physical disadvantages (he was extremely nearsighted); likewise, the rejection of George Ingle Finch for the 1921 British Mount Everest reconnaissance expedition has been represented as a snub to a self-made "colonial" without public school or Oxbridge connections (and has also been blamed for holding back British use of oxygen, as Finch was a keen promoter of gas technology).[8]

These narrow circles of acceptability may have broadened after the Second World War as expeditions became increasingly international. By the 1950s there was a clear pathway for colonials—particularly New Zealanders—to join British expeditions; the next Brit to get to the South Pole (after Scott's team) was Vivian Fuchs, in 1958, as part of the Commonwealth Trans-Antarctic Expedition (TAE), which included a second team led by New Zealander Edmund Hillary. The TAE was also the prompt for a major physiological expedition to Antarctica, combining American, British, and German expertise. But none of this internationalism meant that it was less important to be the "right sort"—indeed, one of the key qualifications for acceptance on an expedition team was the rather circular requirement that one should already have completed an expedition.

This self-feeding mechanism was one of multiple reasons why women were so rare on scientific expeditions: the requirement to have expeditionary experience *and* scientific credentials effectively acted as a double disqualification. There are, of course, exceptions, although they generally prove the rule that scientific expeditions were all male. Perhaps the first is Mabel Purefoy FitzGerald, briefly mentioned in the previous chapter, who as an Oxford-educated scientist (she took courses in chemistry and biology although, of course, she was not allowed to graduate) developed an interest in respiration physiology and collaborated with J. S. Haldane in the early years of the twentieth century.[9] In 1907 she moved to the Rockefeller Institute in New York, on the back of a Rockefeller Traveling Fellowship, to study bacteriology—but when she heard that Haldane would be at Pikes Peak, she contacted him directly and offered to take part in the expedition and help them set up a base at Colorado Springs.[10] The men of the expedition (Haldane and C. G. Douglas of Oxford University, Yandell Henderson of Yale, and Edward C. Schneider of Colorado College) were not happy with the idea of a fifth, female member of the team, so instead, while they researched on Pikes Peak, FitzGerald conducted a solo research trip across mining camps in Colorado, measuring alveolar P_{CO_2} in miners. Her solo-authored paper on this data, plus another on a similar expedition to the Appalachians, are still widely cited, as they were some of the first measurements of the kind to be taken at mid-altitude and not only created a baseline for such figures, but also "fill[ed] an important gap in the spectrum of alveolar gas values from sea-level to extreme altitude."[11]

There does not appear to have been another woman doing physiological work at high altitude until the late 1950s, and even then, women's roles were very effectively concealed by the conventions of scientific publishing. In 1959, as part of the British Sola Khumbu expedition led by Emlyn Jones, the expedition's medical officer, Frederic Jackson, pursued a project to compare the ECGs of permanent residents at high altitude with those of temporary visitors. His work was facilitated by the assistance of Nea Morin, a pioneering mountaineer and rock climber. Morin helped "Fred" set up a surgery, assisted in his examinations, and acted as scribe, taking down the names and ages of the adults and children he examined.[12] And yet, in Jackson's published account of this work, she is not mentioned anywhere, not even in the acknowledgments, except as "one woman" who was merely a *subject* of the examinations.[13]

A woman, albeit unnamed, was also a member of an expedition *and* the subject of a physiological study as part of the sleep studies of the London School of Economics' Mountaineering Club's Himalayan expedition of 1956.[14] Depending on whether one considers data collection of this sort to be sufficient to count as "doing science"—effectively a form of "natural history" in the traditional systematics of scientific practice[15]—either this anonymous woman or the slightly more involved Nea Morin has a good claim to being the first woman to do biomedical research in the Himalaya.[16] Yet these women, too, were exceptions. As late as 1985 John West wrote to defend the decision to deliberately exclude women from the American Medical Research Expedition to Everest (AMREE), not only on the grounds that—as he asserted—there were few women trained in high-altitude physiology and that men were simply stronger climbers, but also because "women might create additional tensions that an expedition of this complexity could ill afford."[17] This statement, and the absence of women in the archival and published records of the AMREE experiment, led me to assume that they were not present until my serendipitous discovery of an article in the *Journal of the Indiana State Medical Association*, written by the doctor Polly Nicely and the nurse Judith K. Childers. It begins as a typical outline of the expedition before revealing, on the second page, that both of the authors traveled with the AMREE team, as part of a fund-raising "support trek," as far as Everest Base Camp.[18] Not only that, but both Nicely and Childers "participated in studies of our own regarding personal health and the effects of altitude. A graph was kept of pulse, respiration and symptoms."[19]

Women fared better in the Arctic, not least because of the huge geographical range of the circumpolar zone and the ability of women to travel via their own countries into the Arctic region. In an article on the founding of the US's flagship Arctic Research Laboratory at Barrow Point in Alaska (est. 1947), women are present in photographs of seminars and in group staff photos (which also show an African American researcher from Swarthmore College, one of the key academic players in the founding of the university).[20] Even where access was more challenging, women were able, as early as the 1950s, not just to work at sites that were extensions of their "day jobs," but also to lead scientific expeditions. Probably the first such expedition in physiology was the Cambridge Spitsbergen Physiological Expedition of 1953. A mix of university staff and students in natural history and physiology, it was led by physiologist Mary Constance Cecile Lobban (1922–1982), and two of the other seven participants were also women (both recorded as "physiographers"). This expedition spent a little over a month (June 27–August 30) on the island "establishing a laboratory at Brucebyen" and conducting studies into the effects of prolonged daylight on renal function.[21] Lobban led a second expedition in 1955, and one of her self-defined "human guinea pigs" described the process as explicitly an *expedition*, not "merely" an *experiment*:

> I was impressed by the vision and determination which brought some dozen subjects to live under a strict regime in the isolation of Bruce's old huts . . . Lobban was very conscious that she was leading an expedition in the tradition of Scott and Shackleton, not merely organizing an experiment.[22]

Lobban subsequently got a job at the National Institute of Medical Research's (NIMR) Division of Human Physiology in the UK and then, in 1978, a professorship in environmental physiology at the Memorial University of Newfoundland; she worked predominantly on circadian rhythms, particularly within the Arctic Circle.

Circadian rhythm research also demonstrates that even though women were rarely allowed into the cliques of extreme physiology, they were still able to make use of the same networks men relied on to consolidate careers in the field.[23] Because many expeditions were multipurpose, women had opportunities in the Arctic to slip between research programs: for example, in 1959, Helen E. Ross took advantage of the Oxford Finnmark Expedition—a primarily ornithological expedition to Arctic Norway—to study sleep patterns of both male and female subjects using cards provided by H. E. Lewis from the MRC. (This study led to the rather unfortunate conclusion for gender equity that "the women slept significantly longer than the men when in the Arctic, but it was not clear whether this was a physical need or due to laziness.")[24] Meanwhile, women continued to do essential "invisible technician" work, fund-raising, designing and packing of instruments, and data analysis—work such as that conducted by Jean Rodahl, discussed in chapter 1.

The Antarctic, however, remained a closed shop to female scientists until the 1970s. This exclusion was partly an active choice on the part of the male scientists and explorers engaged in Antarctic research, and partly achieved through passive or secondhand means, such as the refusal of the US Navy to transport women to Antarctica (making it almost impossible for European or American women to reach the continent). A handful of women had set foot on the continent in the pre-IGY period, including one American in 1947–48, Edith Ronne, who, with Canadian Jennie Darlington, was one of the first women to overwinter there. Both these women were traveling with their husbands, a common pattern for women in extreme environments. Other nationalities were likely to be represented if they had national bases—women from Australia, South Africa, and the USSR were able, at least, to do oceanographic research, if not to explore on land. A few British women came thanks to these bases, too, most notably in 1959, during one of the Australian National Antarctic Research Expeditions (although it traveled to Macquirie Island, which is technically sub-Antarctic),[25] in which Mary Gillham traveled with three other women: the celebrated Australian marine biologists Hope Black (née Macpherson) and Isobel Bennett and the British-born biologist Susan Ingham. In the following year Macpherson and Bennett returned, this time with the botanist Elise Wollaston and the historian Ann Savours (who is cited above as the obituarist for Mary Lobban, the Arctic scientist). Women, just like men, created their own international networks, where they could. By the 1950s, however, there was an official ban on American women on the continent of Antarctica, which was lifted only in 1969 by the US Congress. Women (and some men) did challenge these bans and bars: New Zealand geologist Dawn Rodley, who was also a seasoned hiker and climber, was offered a place on a 1958 IGY-related research trip and argued for her right to travel, but the US Navy refused to help transport her to the continent as part of Operation Deep Freeze, and she was unable to take up her place (this despite the fact that in order to secure it she had had even asked permission of the *wives* of the other expedition members).[26]

British women's relative lack of access to Antarctica was the result of soft pressure rather than hard rules. Vivian Fuchs, director of the Falkland Islands Dependency Survey (FIDS), which later became the British Antarctic Survey (BAS), was explicitly opposed to the presence of women on research bases, believing them to be a disruptive influence.[27] During a shortage of doctors immediately after the Second World War, the Falkland Islands government considered a female doctor for a position on the remote South Georgian base being used by FIDS/BAS, but "that was felt to be ill advised."[28] British women did not get invites to British bases until 1983, and no British woman overwintered there until 1996.[29] In 1965 Ove Wilson (medical officer of the 1949–52 Norwegian-British-Swedish Antarctic Expedition) wrote in a chapter on human adaptation in Antarctica that

> irrespective of the practical disadvantages in accommodation, toilet facilities, privacy, etc., the presence of women would probably have a more disruptive influence on station life than a beneficial psychological effect, judging from experience.[30]

Notably, the "experience" he claimed to be judging from was largely based on his reading of two popular accounts of Ronne and Darlington's overwinter in 1947–48, not actual lived experience of working with women. As Morgan Seag has shown, these appeals to "disruption" or to the very trivial "lack of facilities" were the main arguments used right into the 1980s to resist women's participation in Antarctic research, acting, as she suggests, as acceptable proxies for a more general misogyny in a period when national legislation, such as the UK's 1975 Sex Discrimination Act, made the exclusion of women less acceptable in a legal, as well as a social, context.[31]

Because of these bans, soft and hard, even when women had the scientific credentials and reputation to argue for a place on an expedition, their ability to gain experience working in extreme environments was significantly restricted. "Experience," in a rather chicken-and-egg way, was often discussed as a critical factor in someone's value as a potential team member. On the one hand, experience was a valuable and rare commodity, and those who had it would find themselves called on repeatedly. On the other hand, experience was also a proof against the problem of the last chapter: the difference between models of extreme environments and nature itself. No reliable sea-level test existed to discover whether a scientist was sensitive to altitude, and the best predictor of someone's potential to work at low oxygen pressure was their past performance. Similarly, despite regular discussions about the psychology of expeditions, one did not really know whether an applicant would make a bearable—let alone beneficial—expedition member until one had gone on an expedition with them. Consequently, among letters discussing, critiquing, and recommending gas masks, food rations, and packing firms, there runs a line of correspondence doing exactly the same thing for individuals. So-and-so is a "good chap," "reliable," "the right stuff." Take, as an example, George Lowe, a New Zealand–born mountaineer and schoolteacher. His first high-altitude expedition was to the Himalaya in

1951, where he climbed with Edmund Hillary. Consequently, he was invited on the 1952 expedition to Cho Oyu with Hillary, Shipton, and Pugh, testing oxygen equipment, and then on the 1953 British Mount Everest expedition. In 1954 he and Hillary attempted to summit Makalu, and although their effort failed, Lowe met Vivian Fuchs, who invited him and Hillary to participate in the TAE—despite the fact he had no Antarctic experience.[32]

Expeditions such as these were closed and careful worlds, for good and bad reasons. Virtually all the scientist-explorers named in this chapter, or the previous one, experienced the loss of colleagues, friends, and team members or saw horrendous and life-altering injuries, and they were working in tense, high-cost situations, which in some cases represented once-in-a-lifetime opportunities for work (and play). These situations were not conducive to taking a risk on people. The need for reliability did not restrict itself to the Western participants: where guides were used, most obviously in the high Himalaya, expedition leaders showed a distinct preference for re-recruiting tried-and-tested Sherpa, followed by choosing those recommended by trusted Western climbers. So concerns about picking the right team led to the formation of and perpetuated an extremely restricted research community, including its invisible technicians. Even by the end of the 1970s, when, as David Kaiser has suggested, the "hippies were saving physics,"[33] West, as organizer of the AMREE, expressed concerns about a potential member, Dr. Karl Maret: although Maret had high-altitude experience, a medical qualification, and a master's in bioengineering, which made him perfect on paper for a newly created expeditionary role working with cutting-edge technology, he dressed "like a hippie" and had been found working to make ends meet as a waiter in a vegetarian restaurant, which was enough to cast doubt on his status as a "good chap."[34] Fear of "disruption"— that is, that they would not be "good chaps"—could keep out vegetarians as well as being the reason to exclude women from expeditions, long after they had proved themselves as robust explorers and competent scientists.

One line this anxiety took was expressed as a desire to protect the professionalism of explorer-scientists. The letter quoted above from an ambitious would-be extreme physiologist to Pugh is just one of many similar queries from expedition doctors (including student doctors) asking about potential research opportunities. While Pugh and others responded with advice, there is also a sense in these letters and responses that such "amateur" or ad hoc research programs were thought of as having relatively little value to the "serious" investigation of altitude physiology, or even as being bad for its reputation. By 1956, the NIMR had committed to training all medical officers of the BAS in order that their unique opportunities for physiological, epidemiological, and even clinical research in the Far South would not be wasted, and so that some long-term schemes could be planned and enacted (twenty-five medical officers were trained in the first nine years of this collaboration).[35] The desire for "professionalism" in the sphere of extreme physiology, yet another reflection of the demand for *experience*, is consistent throughout the twentieth century. As late as 1993 John West refused to send equipment for alveolar air sampling to a fellow doctor on the grounds that it was going to be used by the mountaineer Edward Viesturs; West argued that in order for anyone to take measurements that were actually worth having, they needed first to have previous experience of taking such measurements (and of keeping them safe, and analyzing them at home).[36]

These anxieties about creating strong teams and successful scientific projects are, of course, amplified by the genuine life-or-death challenges faced by extreme scientists—whether falling down the Western Cwm or starving to death on the march back from the South Pole. Frequently, very small decisions or events (losing or finding an ice axe, for example) can be the difference between survival and doom. For more than a century, the failure of the *Terra Nova* expedition has been picked over by (would-be) explorers, scientists, and historians in an attempt to learn from experience and to discover exactly what decisions led to the deaths of its members in the Antarctic. As is so often the case, such analyses tell us as much about the era of the writers as they do about the 1910s. The character of Edgar Evans, the first to die, has been scrutinized and his "failure" linked to the fact that he was the only working-class member of the team; conversely, Scott's own leadership has been criticized, particularly as part of a post–First World War class critique that challenged the assumption that wealthy elites and aristocrats were best suited to positions of military and political leadership.[37] Toward the end of the twentieth century, however, Scott was reclaimed as a hero, in part because of the scientific work of his expedition, played up in contrast to Roald Amundsen's unacceptably single-goal, "sports"-oriented expedition.[38] It is certainly a tempting area for counterfactual histories: What if the weather had been better? What if the food supplies had been different? Was it incipient scurvy that made the team so weak?

A century after Scott's death, Mike Stroud—a British explorer and doctor—reviewed "a century of learning about the physiological demands of Antarctica" and concluded that even after a hundred years of research, "there is a perhaps surprisingly limited amount that can be done differently."[39] The one major difference between an expedition in 2011 and one in 1911, he said, is *food*. As the following chapter will discuss in more detail, in Scott's time there was a lack of even basic knowledge about diet and endurance exercise; not even the quantity or proportions of fats, proteins, and carbohydrates necessary for hard work were known with any certainty. While it is likely that the lack of micronutrients such as vitamin C, and especially chronic dehydration, affected the expedition's members, their single biggest problem was that they were starving to death. It was not until Stroud himself analyzed energy use on a re-creation of the South Pole trek, the "Footsteps of Scott" expedition of 1984–86, that it was realized that man-hauling of sledges might demand an intake of 7,000 calories a day, or more—compare this requirement with the actual ration in 1911 of, as Stroud calculated it, around 4,900 calories a day.[40]

Quotidian technologies matter enormously in environments like Antarctica; any flaw in the contents of a ration pack, the reliability of stitching on a glove, the robustness of a gas stove in transit, can cost a finger or a life. Consequently, opportunities to research these technologies are rarely available and keenly pursued. Stroud, for example, got to repeat his physiological studies in Antarctica in 1992–93, when he and British explorer Ranulph Fiennes attempted the first unsupported crossing of the Antarctic continent; that is, they carried all their supplies and equipment with them rather than relying on depots or airdrops. (Their expedition was technically a failure, as they were unable to cross the Ross Ice Shelf to the open sea, but is usually regarded as the first unsupported crossing.) During this expedition Stroud reckoned that their caloric intake rose as high as 10,000 calories a day. This expedition was not the first overland crossing of the continent; a successful *supported* crossing had taken place in 1955–58, and had also been a site of physiological research.

Physiology in the Cold: IMPing and INPHEXAN

After Scott's team retreated from the South Pole in January 1912, they were the last humans to arrive there on foot for another forty-six years, until the arrival of the Commonwealth Trans-Antarctic Expedition (TAE). The TAE was an international attempt to cross the Antarctic continent; two teams, one led by Vivian Fuchs and the other by Edmund Hillary, started from opposite sides of the continent with the aim of meeting at the South Pole. Hillary's team unexpectedly got to the pole sixteen days before Fuchs's crew arrived there (in the original plan, Hillary's team was supposed to turn back after leaving the last depot for the other team), and the two groups met on January 19, 1958. Hillary then evacuated by plane to Scott Base, while Fuchs's team completed the crossing following the trail of supplies left by Hillary (who was later dropped off by plane to join that team partway through its second leg across Antarctica).

These explorers were not alone at the pole: American physiologist William Siri flew in from McMurdo Station to take blood from Hillary and other volunteers. He was also part of yet another international team: the International Physiological Expedition to Antarctica (INPHEXAN). This research group consisted of an "American" contingent of three American physiologists (Siri, Nello Pace, and Charles Meyers) and a recent German immigrant (Gerhard Hildebrand), and a "British" contingent of two UK-based researchers (L. G. C. E. Pugh and James "Jim" Adam). If these lists are beginning to make Antarctica sound crowded, that sensation is not misplaced, as substantial numbers of military and scientific personnel were present there during the International Geophysical Year of 1957–58; or, as Siri put it, "This was the Geophysical Year and so there were geophysicists racing all over the continent."[41]

Although physiological studies had previously been conducted in Antarctica—such as *Terra Nova*'s nutritional studies, which will be discussed in chapter 4—many of them had been incidental work rather than being central to an expedition; they included such activities as surveys of hemoglobin from blood taken as part of routine medical testing (useful to determine whether the disease "polar anaemia" really existed) and a study of the iris color of returning explorers (in the early twentieth century there was a live hypothesis that eye color, as well as skin color, changed in response to the Antarctic environment).[42] INPHEXAN was therefore the first dedicated biomedical expedition to Antarctica. It is also a direct connection point between high-altitude and polar research: led by physiologists who were making their names in altitude studies, it was first imagined during an expedition to Mount Makalu and in turn inspired the Silver Hut expedition, outlined in the previous chapter (and Siri went on to be the physiologist on the AMEE in 1963). Just as networks and cliques of researchers were formed in these extreme environments, they also created a geographical network of activity, as likely to connect Everest Base Camp with a military installation at the edge of the Ross Sea as with an environmental testing chamber in Farnborough or a blood analysis laboratory in California.

The motive force behind INPHEXAN was Nello Pace (1916–1995), an American physiologist credited with promoting a research theory that physiologists needed to study the human body under abnormal, as well as normal, conditions.[43] Pace worked most on altitude physiology, driving the foundation of, and then running, the White Mountain Research Center in California (est. 1950) and taking part in Himalayan expeditions, but he also studied human responses to extreme cold and advised NASA in various capacities relating to the effects of spaceflight on the human body. Pace and Siri had participated in an expedition to Makalu in 1954, the first American expedition to a Nepalese region since the "opening" of the country in 1949.[44] This was not primarily a scientific expedition, but was instead the first attempt to climb Makalu, led by Siri due to his prominence in the Sierra Club rather than because of his scientific interests; Pace went as deputy leader and conducted physiological studies at Base Camp. After the trip (which was unsuccessful in that the summit was not reached), the two men discussed their interests in blood, adaptation, survival, and stress and formulated ideas about the possibility of an expedition to Antarctica. Meanwhile, in the UK, Pugh had also been considering the possibility of research in the Far South—inspired not only by his work on Cho Oyu and Everest, but also by a line of research into thermal physiology, which at the time was manifested in studies of temperature regulation in long-distance swimmers. Pugh looked to his employer, the MRC, for financial support, and initially contacted Fuchs directly to ask about the possibility of taking blood and other samples from TAE team members.[45]

It is not entirely clear from the archives how the two independently imagined physiological programs merged, but in spring 1957 Pace suggested that the two projects should become one and that Pugh should be formally invited on the Antarctic trip, which could now properly be called an *international* physiological expedition.[46] Although nominally based at the US McMurdo Station on Hut Point Peninsula, the INPHEXAN team traveled widely across the continent, sledging or flying to other permanent and semi-permanent stations to collect samples of blood and urine from Antarctic workers (mostly geologists) and, of course, making trips to the South Pole itself. A few days after Siri met with the TAE teams, Pugh, Adam, and Charles Meyers (an employee of the Naval Biological Laboratory, Oakland) took a day trip by plane to the Amundsen-Scott Station at the South Pole to collect data on ultraviolet radiation and microbial samples. As Siri put it, "As a result of these excursions we probably saw more of the Antarctic than most people who have been there."[47]

The reason for these trips was that the team members were particularly interested in studying physiological responses to environmental stress—primarily cold stress. On arriving at McMurdo Station they apparently found the huts heated to such a degree that "there was no possibility of finding someone stressed by cold exposure at the base."[48] Indeed, Siri and Pace immediately "abandoned the overheated hut [they] had been assigned to and moved into a canvas-covered affair that had all the benefits of hot and cold running drafts and snow blowing in through the ill-fitted plywood door, and finally felt comfortable in [their] sleeping bags."[49] The needs of the experimenters also dictated a certain amount of travel around and away from Hut Point, as Pugh wrote to Pace in July 1957:

Those members of the party who are to act as subjects must expect to be on the move out-of-doors for at least 4 hours a day. Otherwise the significant changes will be missed, because temperatures in the tents must necessarily be kept above freezing point while physiological work is in progress.[50]

Despite the cooperative international nature of the expedition (and its human guinea pigs, as around a dozen different countries had geophysicists, meteorologists, and other researchers on the continent), the work of the American and British teams was discrete. Siri, Pace, and Hildebrand were particularly interested in hormonal (adrenocortical) responses to stress, while Pugh and Adam had perhaps a broader remit, doing metabolic studies and focusing on the adaptation of the body's extremities (hands and feet) to cold. Both teams collected, or assisted in the collecting of, body temperatures, subjective experiences of cold, and weather recordings. The internationalism undoubtedly added an administrative burden to the expedition: it took a series of letters over several months between Pace and Pugh, and between the Office for Naval Research (ONR) in the US and the MRC in the UK, to get Pugh and Adam funding, clearance to travel with the American team, and leave from their normal jobs.[51] At the same time, that internationalism was clearly important: while Pace justified it to the ONR and the MRC on the grounds of scientific excellence and international cooperation,[52] Pugh more directly suggested that a mixed team was important to gain the agreement of some crucial human guinea pigs: "I foresee much easier acceptance of our party by British and New Zealand groups if there is adequate British representation."[53] Specifically, Pugh used this argument as a way to justify the presence of Adam:

Allan Rogers, the physiologist to the Trans-Antarctic party, was trained in metabolic methods by Jim Adam, and this would make collaboration easier. I have gathered from scraps of laboratory gossip, that he was not too happy about co-operating with a virtually all American party.[54]

It is difficult to overemphasize how important cooperation was to the researchers in Antarctica. Blood sampling alone clearly alarmed some participants—even hardened explorers like Hillary—and other experiments required extraordinary dedication from participants. One stark difference between the research done at high altitude and in extreme cold is that the complexity of procedures possible in the latter setting is far greater—much more like the sort of physiological work that could be done at mid-altitude. As late as the 1980s barometric chamber studies were, as we have seen in the previous chapter, justified by the limits to study placed on researchers by the environment on Everest. But in Antarctica, more advanced and complex work was possible. One example of such work is the "IMPing" study conducted by Allan Rogers. The IMP, or integrating motor pneumotachograph, is a piece of experimental kit designed to measure energy expenditure in humans. It was developed in the mid-1950s by Dr. Heinz S. Wolff at the NIMR (where Pugh also worked), under the instruction of Otto Edholm, for the purpose of monitoring the metabolic needs of army recruits in order to reform feeding and exercise regimes in the British military services.[55] The core of the machine, a face mask and airflow meter, was connected to electronic monitoring equipment that could be carried by the subject in a backpack. The IMP was, effectively, an adaptation of the Douglas bag, a piece of physiological equipment that collects expired air for analysis, developed by C. G. Douglas while he was working with Haldane in the first decade of twentieth century, but which has been used and adapted extensively by physiologists to the present day.

The aim of the program that led to the IMP was to produce a device that could monitor the metabolism of its subject for twenty-four hours or more, rather than the two or three hours possible with the traditional Douglas bag. As part of the extended INPHEXAN work, many scientists and explorers were recruited by Allan Rogers to wear this equipment that measured, in the words of the explorer George Lowe, "the 'horse-power' rating of a human being."[56] Probably the most intensive work was done on Geoffrey Pratt, a seismologist in the TAE party (and therefore one of the geophysicists "rushing around": see fig. 3). Pratt wore the equipment for seven days straight while he engaged in his scientific work, finding it bulky, awkward, and stressful, while Rogers "dance[d] attendance on Geoffrey day and night, adjusting the 'IMP,' changing it with every change of clothing, sealing off glass air samples, checking 'IMP' performance, weighing every item of Geoff's food."[57] Pratt wrote a report about his experience titled "On Being Imped," which is a detailed breakdown of the challenges of being an experimental subject. From the "difficulty of blowing away sawdust and filings" to the fact that the mask made it impossible to communicate with colleagues, the machine clearly impeded normal practices. But it also presented challenges specific to Antarctica and similar environments that might not have been discovered without direct experience:

> You cannot smell or taste your fingers to see if they are clean. This is necessary because (a) water shortage prevents your assuming they're dirty and washing anyway, and (b) in the course of processing it is necessary to handle the oil heater, which quite often results in kerosene . . . y [sic] fingers.[58]

3. Geoff Pratt, pictured at Depot 700 during the TAE. (© Antarctica New Zealand Pictorial Collection [1957/58].)

Some of the challenges of the IMP were mental as much as physical: as Pratt wrote, "you never, for a single moment, escape from a suffocating feeling and a very conscious effort in breathing." Pratt lists a series of small errors or moments of clumsiness (having to repeat calculations, breaking a compass, spilling ink) for which he blames his IMP.[59] (Pratt had a particularly unfortunate expedition in terms of respiration, as a year later he was found to be suffering from acute carbon monoxide poisoning as a result of his research and had to be given oxygen from the team's welding supplies until medical O_2 could be airdropped by the US Navy.)[60] The IMPing project also highlights the reality of interdisciplinary cooperation; expeditions could not afford to have single-subject specialists, and in many cases scientific professionals juggled dual roles—as team doctor and research physiologist, or as geologist and official photographer. As Rogers put it in a later report, "The logistics of the crossing were so critical that no passengers could be carried."[61] Less formal cooperation was common, too—weather readings were crucial for the studies on thermal stress, so there are frequent references in the researchers' notes to the taking of maximum and minimum temperatures as well as other, strictly speaking, meteorological measurements by non-meteorologists. Not all of these tasks were straightforward: on January 4, 1958, Pugh recorded in his diary that a cover blew off while he was trying to take [solar] radiation readings and he had to "chase it nearly half a mile over dangerous ice."[62] Rogers noted that on the TAE, "everyone helped with the meteorological records" even though they were, officially, "the entire responsibility of one man, Hannes La Grange."[63] In return for Pratt's service as a guinea pig, Allan Rogers ended up working as a "geologist's mate," assisting in the seismological studies, carrying rocks, and fetching tools.

Awkward Data on Acclimatization

The TAE and INPHEXAN scientists generated a huge amount of data, both written (as transcriptions of temperatures, calories, measurements, sensations) and material (as vats of urine, vials of blood, cores of ice). While all scientific data is subject to travel and transformation, the material produced by the INPHEXAN team, which is largely representative of the data gathered on any expedition discussed here, underwent one of three immediate fates: it was analyzed in situ, it was distributed across global networks, or it was transformed for storage in the hopes of later analysis. The last category includes the meteorological data, which was really used only in connection with other forms of evidence. For the physiologists, it had value when it acted as a control factor in other analyses—of body temperatures, clothing preferences, hormone levels—while for the other scientists "rushing about," the weather data was, of course, their primary purpose for being in Antarctica.

The facilities at the various Antarctic bases allowed for some analysis to be conducted immediately—for example, by the late 1960s the American base at Plateau Station was well enough equipped for frequent use of basic blood analysis (hemoglobin concentrations, white blood cell counts, and similar tests) and urinalysis (usually measurements of blood, protein, and glucose levels and pH measurement).[64] Blood and urine were also processed in situ during the INPHEXAN project.[65] These measurements were often taken in aid of physiological research, but they were also important as diagnostic tools for the medical officers at various bases. At the same time, samples, including blood and urine but also more "exotic" substances such as tooth plaque scrapings, were exported from Antarctica to laboratories across the world.[66] These samples were generally sent back to "home" laboratories: "hundreds of such samples" of blood and urine collected during INPHEXAN were sent back to Berkeley alone.[67] In some other cases, however, materials were distributed more widely: in addition to his human studies, Pugh caught and killed seals to examine their blood, but because this study had not formed part of his original experimental design, he drew up his plans to analyze the animals' physiology only once on the ice. He analyzed some of the blood and fat in situ for carbon monoxide levels, but he also arranged to send prepared seal blood samples to the Pathology Department at Christchurch Hospital in New Zealand.[68] The result was not entirely satisfactory: the thawing of the samples created a vacuum powerful enough to crack the container and allow air in. In a panic, the researchers in Christchurch removed the plasma with a syringe, only to discover it had only partially thawed, so the material they ended up analyzing was condensed plasma with the water frozen out and therefore gave distortedly high readings for carbon dioxide concentration.

The difficulty of transporting samples from precarious research sites was discussed openly in extreme physiology publications. As early as 1924 Howard Somervell had attempted to take alveolar air samples on Everest, at altitudes as high as 23,000 feet (7,010 m) above sea level.[69] These samples were sent down to Base Camp for analysis, but later researchers have claimed that his results were not representative because he used rubber bags—in fact, football bladders—for his samples, which allowed the diffusion of carbon dioxide. Trying to improve on this method, Raymond Greene, who took samples on the 1933 British Mount Everest expedition, decided that analysis in the Himalaya was impractical, and instead sealed the air samples in glass tubes and shipped them back to the UK.[70] Pugh used a method similar to Greene's to collect air samples in both 1952 (on Cho Oyu) and 1953 (on Everest), and his effort also illustrates the importance of field-based adjustments and learning from experience: in 1952 Pugh sealed his glass ampules using a flame from a Primus stove, but finding this method "difficult and uncertain," he ensured that, during the 1953 Everest expedition, he had a specially designed "cylinder of butane fitted with a Bunsen burner," which was taken as high as 24,000 feet (7,315 m) above sea level.[71] Pugh's samples from both expeditions were also circulated; some were analyzed on Everest itself, and others were sent back to London for processing.

By the time of the AMREE expedition in 1981, the technology for sampling had become much more efficient and automated, although even then the environment of Everest proved challenging. Although he was equipped with an automatic air sampler, Chris Pizzo found that it had frozen and would not rotate properly on the summit, requiring him to put in the ampules by hand: "It was hard. Those fourth, fifth, and sixth ones were really hard, and I remember the exhalation . . . was making me a little dizzy."[72] These samples were carried, personally, by John West in a "special box" back to UCSD for analysis.[73] Such personalized transport was still necessary, as unpredictable conditions could ruin samples otherwise—as was the case for some of Pugh's hard-won seal blood. It was also the case for blood samples taken on Kanchenjunga by a British and Indian team in 1954: although the mountainside samples proved reliable and easy to analyze, the "control" samples taken in Darjeeling and flown to Calcutta for analysis were "haemolysed or decomposed," almost certainly because of an unpredicted heat wave in the area (in which the maximum temperature in the shade was over 43°C).[74]

Other studies required other body materials. The rather incongruously named "Operation Snuffles," a US Navy research program that ran in 1958–59, took both blood and "swabs" (presumably of mouth or nose) from men stationed on the USS *Staten Island* and at the Cape Hallett and Wilkes Stations in Antarctica.[75] This study produced an astonishing 900 blood samples, 1,300 viral cultures, and 2,660 bacterial cultures, which were sent on to "low-temperature freezers in Baltimore" for distribution to research centers: the viruses and sera to the National Institutes of Health and the bacteria to Johns Hopkins University.[76] Less expansive, but more unusual, was a study in the 1970s, as part of the IBP, on dental decay, which limited a group of nineteen British Antarctic workers to stores whose food contained sucrose for six months of the year, and to an artificial sweetener plus glucose syrup for the other six months. For this study, each participant's tooth plaque was carefully scraped off, frozen at −20°C, and eventually shipped back to the UK for analysis.[77] (When considering the cost of human experimentation, it should be noted that these collections happened once every two weeks, and that the participants were required to "refrain from oral hygiene" for three days before the collection period.) Urine was also used to analyze dietary responses: in 1969 the British researcher R. M. Lloyd collected his own and ten colleagues' urine during a year-long stint at Halley Research Station in the Antarctic. This material was also frozen (again at −20°C) and also shipped back to the UK, to Liverpool, for ketone body analysis, whose results suggested that, whether on high-fat or low-fat diets, men hauling sledges in the Antarctic do not acclimatize sufficiently to respond to the heavy metabolic demand of hard work in extreme cold.[78] Similarly, in analyses of samples of body fat clipped from the buttocks of unfortunate British Antarctic Survey workers and sent from Halley Research Station to Johannesburg in 1962, no evidence was found that human workers put on body fat in response to cold stress, in contrast to findings in laboratory tests of animals.[79]

If data collected in the Far South was not analyzed locally or at a distance, its final fate was to be converted into new forms to await future analysis. The TAE provides a perfect example of such conversion—involving possibly one of the longest delays between the return of data and its publication—as the information gathered about clothing and adaptation took thirteen years to be converted from field data to printed paper. The data was gathered as part of a non-IMP-based physiological project led by Allan Rogers, which took advantage of his role as medical officer on the TAE. His study was an attempt to investigate one of the ongoing physiological puzzles of extreme cold: Did human bodies adapt to low temperatures? Clear evidence had emerged of adaptation to hot, dry, humid, and high-altitude conditions, but the existence of cold adaptation proved more controversial, as we saw above in the failure of later studies to find adaptation in samples of body fat or urine. To answer the question, Rogers designed a deceptively simple experiment: he asked participants in the TAE to record their clothing over the almost fifteen months they spent in Antarctica. He hoped that these records, which would be kept both at the (heated) stations and on the overland crossing, would show, when correlated with weather conditions, whether men who had lived for a long time in the cold wore less clothing than those who had recently arrived in Antarctica—in other words, whether or not they had acclimatized to their new environmental conditions. The explorers filled in cards recording their daily outfits, their activity and sleeping patterns, and additional data on their medical conditions, and the weather was recorded to correlate with the outfits worn.

The reason this experiment is "deceptively simple" is that although the method seems straightforward, the data it produced was vast. Daily, usually multiple daily, records of the clothing layers on all major parts of the body for about a dozen men, over the course of about sixty weeks, had to be correlated with hundreds of records of meteorological conditions and thousands of records of activity and medical conditions. Further, estimating the thermal value of the clothing was itself a challenge—after all, is a man wearing two shirts and two pairs of gloves more or less thermally insulated than one wearing two pairs of trousers and two hats instead? This distinction mattered, as there was some evidence that the extremities of the human body adapted to extreme cold, even if the whole body did not, so changes in layers on the hands, feet, and head needed to be analyzable separately from the "whole outfit."[80] To assess thermal value, twenty-four "clothing assemblies," consisting of the warmest and coolest outfits worn by each of the twelve men studied, were sent to Wright-Patterson Air Force Base in the US, where they were analyzed using a "thermal copper man" (a heated mannequin exposed to different temperatures so that the energy required to keep the "body" at a certain temperature could be measured—for more on these techniques, see chapter 4). For calibration, four assemblies were also sent to the RAF Institute of Aviation Medicine in Farnborough, UK.[81] The analyses at Wright-Patterson and Farnborough produced thermal values for the outfits, but the two laboratories did not end up with the same values for all the clothing assemblies. Insulation is a very complex property of clothing and can be significantly altered by simple changes, as when a shirt becomes untucked or clothing becomes damp due to snow or sweat.

Despite these problems, the data was collated for comparison, but it proved overwhelming for the first three statisticians who volunteered for the job, all of whom dropped out of the project less than a year after agreeing to work on the numbers. It was not until 1968, when a grant from the US Air Force allowed a full-time statistician to be hired, that the report cards, thermal values, and meteorological observations could be brought together. The statistician hired was a woman, Mrs. R. J. Sutherland, a recent graduate in mathematics who not only dealt with the statistical correlation of the pile of TAE data, but also designed the computer program necessary to run the calculations. Two other women assisted in turning the mass of cards, printouts, and weather notebooks into a report on human acclimatization: Mrs. E. Fountains and Miss V. Thornton were hired (sequentially) as administrative assistants. The latter took on the not insignificant task of coding the clothing cards—in other words, turning the explorers' notes into a classified system of information that could be input into a computer. The final report, jointly authored by Sutherland and Rogers and titled *Antarctic Climate, Clothing and Acclimatisation*, was published in March 1971.[82] Eventually funded by both the European Office of Aerospace Research and the US Air Force, the report concluded that there was no evidence for individual human adaptation to cold: "The results show quite clearly that in this particular survey of this particular expedition, there is no evidence of less clothing being worn to meet the same cold stress as time passed."[83] While this was one of the most detailed studies of acclimatization up to that time, its findings merely echoed what dozens of previous studies had already concluded.[84]

The lack of evidence of short- and medium-term human adaptation to low temperatures may explain why research into extreme cold diverged from the pattern of high-altitude research. In the latter, the long-term, hereditary adaptations of indigenous peoples to low P_{O_2} were at least somewhat related to the short-term physiological responses of visitors, and as such could be part of a single, coordinated biomedical research theme. For low-temperature research, the relationship between physiological studies of indigenous populations and explorers' needs was less clear; if adaptations to cold were cultural, social, or technological, rather than biological, it was harder to see a path from the physiological study of indigenous peoples in the Far North to the bodies of Europeans, Americans, Australians, and New Zealanders in the Far South[85]—at least, that is, from the *racialized* physiological study of indigenous peoples. One of the confounding factors for cold-weather acclimatization studies was the fact that in most situations, Arctic and Antarctic explorers are only periodically exposed to very low temperatures. After all, the whole point of Arctic huts and fur clothing is to keep human bodies at a reasonable temperature, and it quickly became obvious—as was briefly mentioned above in relation to INPHEXAN—that attention needed to be paid to whether or not Arctic and Antarctic explorers really were experiencing "cold stress" at all. This anxiety extended to studies of indigenous peoples, as in a 1949–50 study of cold acclimatization in the "Eskimo" in which the researchers were at pains to identify participants living a "traditional" lifestyle (presumably with more cold stress than a "modern" lifestyle), thereby clearly indicating their assumption that relative immunity to the effects of cold was an acquired, not a racial or ethnic, characteristic.[86] Perhaps if Western explorers wanted to learn from indigenous peoples, then ethnographic, sociological, or technological studies would be more relevant than physiology?

Some of these ideas are explored further in the following two chapters. When the IBP was founded—in emulation of the IGY—in 1964, a "Human Adaptability" theme was added to the program, under which most studies of extreme physiology fell. Although Antarctica was included in the geographical range of this program, and hundreds of studies into adaptability were funded, the IBP was much slower to inspire physiological research in Antarctica than, slightly ironically, the IGY had been. Some international organizations did respond to calls for work in the Antarctic: in 1962 the Scientific Committee on Antarctic Research (SCAR) held a Working Group on Biology meeting in Paris at which quite a few papers on human physiology were presented, yet it was another decade before a full SCAR-backed symposium on human polar biology could be organized, as a collaboration with the International Union of Physiological Sciences and the International Union of Biological Sciences. The presenters at this symposium, held at the Scott Polar Research Institute in Cambridge, UK, in September 1972, represented a good sample of the nations that had worked in the Arctic or Antarctic, including Australia, Canada, France, Japan, and the USSR as well as the UK and USA (indigenous peoples, however, remained subjects rather than authors).[87] All the nations represented seemed to share the same basic narrative about polar biomedical science: at first it was limited, part of larger expeditions but not their main focus, and mostly conducted by the doctors attending an expeditionary team. Later, it was argued (and this transition seems to have happened in the 1950s for most nations, often in relation to the IGY), research began to involve a more diverse set of professionals, including physiologists and later geneticists, and was conducted within more expansive and long-term research programs.

In contrast to Pugh's Silver Hut expedition, which formed a touchstone for future altitude research, INPHEXAN did not achieve the same significance in extreme physiology circles—it is not mentioned once in the collected papers of the SCAR meeting in 1962 (or '72). It is not obvious why this should be the case, and it is always possible that personal factors may be at play in the ways that expeditions are or are not remembered,[88] but it is a fact that the project led to few publications. The repeated generally negative findings of studies on short-term acclimatization seem to have shifted the focus of polar physiology, as the papers presented in 1962 and again in 1972 show a clear divergence of interests. Adaptation and acclimatization to cold becomes a minority topic, and there is instead a trend toward the study of circadian rhythms, sleep, and the psychology of exploration, in addition to an ongoing interest in nutrition, infectious disease, and the epidemiology of medical incidents at polar stations. Rather than *cold* being a consistent factor, and therefore a key shared interest for physiologists—as *altitude* was for mountaineering research—moderated daylight rhythms and isolation appeared to be the crucial research questions by the early 1960s; research into cold shifted into research into "comfort," "stress," and even heat production as psychological and behavioral factors came to the fore. Meanwhile, food and nutrition remained a key obsession for those traveling to both the Far South and the Far North.[89]

Human physiology played a second-tier role in Antarctic biological research through the twentieth century; SCAR did not create a permanent Working Group on Human Biology and Medicine until 1974.[90] And it was not until 1977 that a second fully international scientific expedition went to Antarctica with the explicit purpose of conducting biomedical research—twenty years after INPHEXAN. This was the International Biomedical Expedition to the Antarctic (IBEA), which ran from 1980 to 1981 under the leadership of the French physiologist and Antarctic explorer Dr. Jean Rivolier, the head of the SCAR human biology group. The IBEA had an expeditionary crew of twelve men, representing five nationalities (four English, three Australians, three French, one New Zealander, and one Argentine). Perhaps paradoxically, it devoted considerable time to the study of adaptation to cold, despite the disappointing findings of decades of previous research. Because its research remit was holistic—"to study the process and extent of human responses in coping and adapting to the conditions of life during an Antarctic journey"—temperature adaptation remained a focus of its research. The participants met in Australia before departure, where half the team were subjected to hour-long cold baths for ten days in a row in a study intended to find out if explorers could be "pre-adapted" to cold.[91] The results of the expedition clearly showed that explorers submerged in cold baths prior to an expedition were no better equipped to handle the cold of an Antarctic trek than those who had not been bathed; in fact, such experiments created fractious relationships between team members—"Did the cold baths have any effects? Most obviously, the men got fed up."[92] In the end, the major findings of the IBEA seem to relate more to the psychology of isolation and teamwork than to the effects of cold climates on the homeostatic regulation of the human body.

The IBEA's research plan rested on the assumption that it would be possible to gather "accurate descriptions of the physiological status of the participants before leaving for the field" to allow for comparisons with their status during their Antarctic trek as well as upon their return.[93] In addition to the cold bath experiments, participants took part in hard exercise in hot climate chambers and in other experiments that "involved the subcutaneous injection of toxins and the infusion of noradrenalin [sic]."[94] Pre-expedition screening also included group therapy sessions—recorded on camera—in which participants were encouraged to talk about their emotional needs and reactions to the expedition. These sessions were not successful because, as one of the expedition's psychologists complained:

> With very few exceptions, the subjects were not sufficiently skilled and experienced in human relationships to be reflective, expressive and supportive of each other. As was only to be expected in the circumstances, the feelings of unfamiliarity, defensiveness and restriction then switched to criticism of psychology as an academic discipline.[95]

Networking with "Good Chaps," Living and Dead

The challenging conditions of the polar regions are unavoidable; the demands of biomedical researchers there are not. With the lists here of cold baths, painful injections, demands for samples of body fluids, requests for participants to take time out to fill in questionnaires (struggling, often, with frozen pens and stiff fingers) and discuss their deepest emotional states with strangers, it may seem difficult to understand why anyone volunteered for such work. But although the demands on human subjects might seem extreme, the lure of extreme environments, especially the "big names" like Everest or the South Pole, could still generate a huge pool of potential volunteers, who had to be screened for suitability. While the importance, and difficulty, of choosing "the right stuff" has always been part of discussions about expeditions, no consensus—scientific or otherwise—has been reached about making such decisions objectively. Military forces did begin to screen men, mostly from around the mid-1950s and mostly using navy psychiatrists (who were experienced in picking men for submarine duty—seen as similarly likely to cause disturbance due to isolation and "cabin fever").[96] But these screens had variable success, and explorers and expeditionary scientists tended to use more subjective methodologies. As William Siri wrote about choosing men:

> The man's skill and experience are also important, although lack of experience wouldn't necessarily rule out a person who seems otherwise to be able, who you believe can learn fast and has good judgment. But all of these elements are involved—confidence in the man's drive, his skills, his experience, his performance under stress, his ability to work with others—all of these things we in some mysterious way try to assess. And you're not always right. People are often great surprises.[97]

(Siri had been involved in the AMEE in 1963, which included a clinical psychologist, James T. Lester, who undertook extensive psychological modeling both on and off the mountain, although it did not reveal anything particularly revelatory about the sorts of men who succeed in mountaineering, or the ways in which they respond to tedium and disaster.)[98]

Some explorers and physiologists had a more sanguine approach to team selection: the Australian researcher Walter Victor Macfarlane said that "conversations with Pugh and others [make it] clear that the very selection of people who can do the job, provides personalities that are fairly compatible with the environment."[99] Macfarlane himself commented that "mountaineering types" seemed to fare well almost by instinct when trapped in mid-altitude huts by unexpected weather in New Zealand. This conclusion assumes that man (specifically *men*) can become acclimatized or resistant to the challenges of the environments in which he finds himself. Whether this process automatically makes a man adapted to his social environment was another matter entirely, and expedition leaders used a range of assessment tools to try to ensure that a team would be able to work together, even at times of shortage and stress. INPHEXAN appears to have been a relatively peaceable expedition—although there were dramas on the TAE, such as life-threatening experiences of planes attempting to land in a whiteout (during Hillary's retreat from the South Pole),[100] an extremely serious case of carbon monoxide poisoning (Geoff Pratt), and potentially deadly sleepwalking (the accident-prone Geoff Pratt, again).[101] In comparison, twenty years later, the IBEA reported much higher levels of dissatisfaction and conflict, including one participant of the twelve who had to be returned to Sydney early due to "homesickness." But the difference between the two expeditions may well be because the IBEA was set up to observe, and therefore be aware and self-reflective about, the psychosocial aspects of expeditions, while INPHEXAN focused on physiology. There is evidence that serious conflicts in other extreme physiology research programs were kept private rather than finding their way into print (see, for example, Pugh and Hilary during the Silver Hut expedition).[102] Additionally, the IBEA team were kept consistently in one another's company, while INPHEXAN researchers worked alone or in small teams, traveling widely about the continent and spending time with the comparatively large IGY workforce in Antarctica.

The expanded population of Antarctica was not to everyone's satisfaction. As early as 1957 Vivian Fuchs was complaining about the new residents of the continent and expressing nostalgia for the "good old days" of polar exploration. After a colleague reported that an American radio operator at McMurdo Station "couldn't place" Ellsworth Station, Fuchs noted in his diary, "So much for the interest of the US personnel in even their own Antarctic activities"—a lack of interest he put down to drafting practices in the navy, which did not prioritize staff with an interest in the continent and operated "without any consideration of the individuals [sic] likes or dislikes, interests or capabilities—what a pity—and how different from the spirit of the old expeditions, or for that matter our own of today."[103] Such random military drafting clearly contrasted with the hyper-careful and cautious selection of "good chaps" for expeditions, scientific or otherwise, and even the patriotic Fuchs had to admit that similar practices were creeping into the staffing of the Falkland Islands Dependency Survey, creating the "same unsatisfactory disinterestedness."[104]

"Disinterested" military personnel did not necessarily make physiological work more difficult; as we have seen with previous studies, such as Operation Everest, it was sometimes easier to get military personnel to submit to painful, stressful, or inconvenient studies than to recruit civilians. In the wartime decade before the IGY, American personnel in particular had been mobilized for extreme physiology studies, although these studies were focused on extremes of heat rather than cold—military personnel marched through deserts or the high humidity of Florida, or floated without water on boats to generate information about heat tolerance and dehydration— creating information that was later mobilized in the desert conflicts in North Africa and elsewhere.[105] But Fuchs's complaint is more about enthusiasm than compliance. The fact that the radio operator specifically failed to recognize the name "Ellsworth" is also significant, as Ellsworth Station (like both Lake Ellsworth and the Ellsworth Mountains) was named after the American polar explorer Lincoln Ellsworth (1880–1951), who, inspired by his earlier attempt to fly to the North Pole alongside Roald Amundsen, became the first man (with Herbert Hollick-Kenyon) to complete a trans-Antarctic flight, in 1935. Failing to recognize the station suggests not only ignorance of the continent's *geography* but, worse, of the continent's *history*, and specifically of the heroes of exploration who had gone before.

Explorers in extreme environments often show an interest in forms of intellectual and academic genealogy, demonstrating a fascination with precedents and previous travelers. The inherently progressive representation of contemporary science predisposes its practitioners to thinking of themselves as part of a "family" of researchers, as do the practices of literature review, citation, and training, which clearly place scientists in a kind of scientific lineage. While scientists may stand on their predecessors' shoulders, explorers often literally tread in their footsteps; the limitations of the human body mean that there are often a finite number of feasible routes over, under, or through a natural space. As discussed in the previous chapter with respect to Silver Hut, this situation can lead to practices such as "pilgrimages" to sites of extreme science, and, as we will see in later chapters, nostalgia can shape discussions of and use of technology. Less obviously, the sharing of physical but not temporal space with previous explorers has materially affected the practice of expeditions and expeditionary science. Sometimes, as we saw at the start of this chapter, it has saved lives.

Part of the network of information created by extreme physiologists was information about abandoned supportive and survival technology. In 1952 Pugh met with the returning Swiss Everest team to use photographs and descriptions to figure out where they had left oxygen cylinders on the mountain. Lowe commented on the useful discovery of a "pile of strong Swiss line" around Camp VII.[106] Swiss caches of cheese, Vita-Wheat, and other "luxuries" were a "sharp relief" from the monotony of the British climbers' high-altitude ration packs in 1953;[107] they feature notably in several autobiographical accounts of the climb, including that by Lowe, who writes fondly of the discovery of honey, "saucison," and a tin of Australian pears: "We ate and spread honey with gloves on and you can imagine what a messy business it was."[108] Swiss remnants also functioned as a reminder of the mutability and change-ability of the mountain: as the British team members made their way through an infamously dangerous part of the Khumbu Icefall known as Hell Fire Alley, they "caught sight every now and then of . . . a Swiss flag, far out to the left of some vast, impossible block over a chasm, [which] would show how much the glacier had changed over the winter."[109] Not only the movement of glaciers, but long-term changes in climate, can be marked with human detritus: the first major attempt to clean up the summit area of Everest in 2010 emphasized that some rubbish from older expeditions was being exposed as the snow line moved and glaciers melted. (It is notable that responses to human detritus shifted from romantic or nostalgic views to frustration and disgust at about the same time the sites became more populated—in the Antarctic and at high altitude, around the 1970s.)[110]

Reuse was not limited to Western explorers: in his account of the Everest and Trans-Antarctic expeditions of the 1950s, Lowe described the Sherpa that lined up for selection in 1953 as

a gay, ragged lot who look anything but the tough and heroic characters they have become known to be. Most of them wear the clothing of previous expeditions: scarlet Japanese silk, saffron cotton, turquoise blue English nylon, multi-coloured wool.[111]

Obviously, the use of climbing gear from previous expeditions was a practical move for people with limited financial resources, who often took clothing, equipment, and boots as partial payment for their services; but it also had a symbolic function as a stark visual cue demonstrating that the man in question had a *pedigree*, experience, which, as we have seen above, was considered a key element of the selection procedure for Sahib and Sherpa alike (although, as we shall see in the following chapter, Sherpa were not always credited with ingenuity for this recycling).[112]

There are, clearly, emotional responses to the discovered and recovered technology and survival materials of earlier expeditions. Returning to, recovering, or reusing historic equipment and sites is a way of claiming a scientific inheritance, part of the desire to be a member of a "family" of researchers. It is a desire felt more broadly, beyond the exploring or scientific community, as evidenced by national preservation societies forming around the wish to preserve and commemorate "extreme archaeological sites" (although this intent does not cover all such sites; for example, those in the Arctic may be paid more attention by tour groups than by historic preservation professionals).[113] There is, of course, a tension between the desire to preserve and the desire to participate in such heritage: in 1957 the journalist Noel Barber and the veteran explorer Sir Henry Wilkins shared a strange dinner of food more than forty-five years old. Consisting of mutton, biscuits, and Stilton, followed by more biscuits and marmalade, the dinner had been dug up by Wilkins from a cache left by Scott in 1910, which also included "scores of tins of English vegetables, some wonderful greengage jam . . . boxes of Quaker Oats, Cerebos salt and Coleman's mustard."[114] While "historic" stores seem so munificent, there may be less urgency to keep them untouched, though the volume of these discoveries emphasizes that even in such isolated spaces humans are never far from the traces of other humans. As Collis has shown with her work on the "footsteps" of explorers' expeditions, reenactments are ways to claim not only a form of *personal* lineage, but also a national or even racial one (in the case of Australia, in direct response to the idea that the land being trodden might belong to someone else).[115] We should also consider a more pragmatic justification for an interest in the past: given that sporting and scientific heroes and heroic journeys carry such popular interest, it may seem to fund-raisers that a connection to a "big name" might help a modern expedition's cause—"Footsteps of Scott" has a ring to it that the "Fuchs-Stroud Overland Metabolic Experiment" does not.

Networking from the Poles to the "Third Pole"

This chapter has sought to outline the ways in which the field of extreme physiology has been created and its boundaries sometimes frozen—how its access restrictions, demands, and cliques allowed some bodies to enter and excluded others. But while its networks were in many ways limited, they were also broad, unexpected, and diverse; they spanned the globe, blurred distinctions between civilian and military work, were frequently international in makeup, and, finally, connected unlikely sites. When Vivian Fuchs invited George Lowe to participate in the TAE, Lowe was being chosen on the grounds of being a reliable "good chap" in the context of *mountaineering* expeditions—not because he had proved his worth or, indeed, had any experience at all, in the Far South. There is therefore an artificiality to the connections between high and cold spaces—they are inventions of the human mind, rather than reflecting an objective "natural" relationship between very different geographical regions. Aside from the fact that the environmental pressures and expeditionary challenges of high-altitude and polar travel are quite different, the infrastructure of such expeditions varies markedly. Antarctica has been, in a significant way and for much of the twentieth century, more isolated than most of the common sites for high-altitude research, and it enforces quite different restrictions on supplies and logistics. Altitude research usually involves engagement with, support from, or experiments on indigenous populations, which are, of course, strikingly absent in Antarctica. For much of the twentieth century, one needed military transport to gain access to Antarctica, whereas it was quite possible to get to high altitude or the Arctic as a civilian tourist (which is one reason for the uneven distribution of female scientists in extreme environments).

It is also not the case that explorers who excel in one environment will necessarily excel in another; there is at least one famous contemporary polar explorer who struggles with acclimatizing to altitude.[116] There are also specific physiological and biomedical differences in responses to these extreme environments, which often combine with accessibility and logistical differences to create divergent approaches to organizing expeditions. The obvious example is diet: at high altitude, flavor sensations and appetite can be significantly altered, and mountaineers often find the higher-fat diets and dietary products of the polar explorer unpalatable, to the point that one of the major problems of high-altitude expeditions in the early twentieth century was, as with Arctic and Antarctic expeditions, starvation, although for quite different reasons—loss of appetite rather than inadequacy of rations. Food rations for mountaineering therefore tended to be more diverse, and higher in carbohydrates and sugars, than polar rations and could be supplemented during the trek in with fresh foods, an option not readily available in the Antarctic away from the penguins and seals of the shoreline.

What really connects these two environments, then, is humans—their motivations and specific interests. Because of the need to take reliable "good chaps" on expeditions, once everyone who *has* traveled to the South Pole is dead, injured, or unavailable, it is logical to turn to those who have survived and thrived in other extreme environments as "good bets" for journeys South. But another connection is the practice of science; what links the work done in these environments (at least at first) is a pressing interest in human limits: how oxygen-deprived, how hot, how cold, how exhausted, can the human body become, and, crucially, how can those limits be extended? Both INPHEXAN and the respiratory studies described in the previous chapter were explicitly about such physical stress and its alleviation. This work carries with it an ethical burden, which will be further discussed in chapter 6, but even without that extended discussion, it is clear how appealing it must be to work at a site where the stress is being caused by "nature" rather than imposed by man, and where one can find human guinea pigs so grateful

4. A view of the summit of the Jungfrau, Switzerland, with the Sphinx Observatory (est. 1937) in the background. This area has been used by multiple Himalayan expeditionary teams—including the Germans in the 1930s and the British in the 1950s—to test equipment and climbers and to do physiological experiments in preparation for expeditions that were to go much higher. (© Ben Allen [2010].)

to be given the chance to experience this environment that they will submit, even at the South Pole, to the terrors of a "10-millliter syringe and . . . lengthy needles" that could cause "even Sir Edmund Hillary's courage . . . to evaporate."[117] That this "laboratory" still failed to give straightforward answers to research questions must have been frustrating for the participants. INPHEXAN produced almost nothing in the way of published papers—even the normally productive Pugh wrote only two items relating to the expedition, and neither was about human physiology (one was a note on carbon monoxide poisoning and the other his observations on seal blood).[118] INPHEXAN's scanty and often unpublished results, and the related findings by Rogers on acclimatization that took thirteen years to be analyzed, consequently do not seem to have significantly affected future cold physiology research, although significant progress was made in some of the other areas described here, such as the estimation of metabolic work done on Antarctic stations, and in areas not discussed in this chapter, such as the treatment of frostbite. Of course, one consequence of the lack of *experimental* certainty was a reinforcing of the value of *experiential* certainty: one could say that an expedition that did *x* succeeded while one that did *y* failed; that when one wore this clothing one was warm, while in that clothing one was cold, even in the absence of "scientific" certainty.

All this networking, this creating of trust and authority, has also developed unusual forms of local knowledge: while the sites of extreme physiology research could be connected by the men and materials that moved between them, their specific demands and unique requirements were also part of the local experience that made someone a valid member of the exploration community. Such experience could be personal, but it could also be claimed by proxy by learning from past generations—including the dead. Indeed, the rediscovery of lost human remains can in and of itself be the motivator for expeditions. The 1999 Mallory and Irvine Research Expedition, for example, explicitly set out to find the body of a "foreign mountaineer" spotted in 1986 by Chinese climber Wang Hungbao (who was killed in an avalanche just days after making the observation, and so could not be asked to pin down the location more precisely). Sponsored by two television companies (the BBC and NOVA), the expedition successfully found a body later identified as George Mallory, and it indirectly enabled his clothing to be analyzed and declared, seventy-five years after his death, to be high-tech enough to withstand an Everest summit attempt.

Other searches for the dead have taken even longer: it was not until 2014 that Sir John Franklin's lost ship HMS *Erebus* was confirmed found, and not until 2016 that his second ship, HMS *Terror*, was officially located, 168 years after the first search party set off to look for the missing Arctic explorer and his crew. The "discovery" of these two ships also reignited a long-running dispute about the attitudes of European and Euro-Canadian scientists toward the knowledge of indigenous peoples. Representatives of the Inuit community in Nunavut, where *Terror* was sunk, argued that the discovery of the wreck was a validation of local oral history, which has a tale of two British ships in distress in the local area and is believed to originate from the middle of the nineteenth century.

In fact, many searchers—in the nineteenth and the twentieth centuries— did use Inuit knowledge to guide their hunt for Franklin, his party, and his ships, and this knowledge was not always helpful or accurate; on the other hand, it was not always taken as seriously as it might have been. In particular, in 1854, early in the search, the Scottish explorer John Rae interviewed local people and uncovered an account of the Franklin tragedy that ended in cannibalism. When reported back in the UK, this scandalous story was roundly denied as a gross slander originating from immoral and backward natives. It was not until the late twentieth century that European and North

American explorers confirmed this story with forensic analysis.[119] "Local knowledge," like experience, is a valuable commodity in extreme physiology work, although its value has sometimes been shaped by prejudices about who counts as a local and whose knowledge can be considered scientific, trustworthy, or rational. The following chapter considers the use of local knowledge, starting with something slightly more palatable than human flesh: pemmican.

FOUR

Local Knowledge

In July 1911 three British men—a doctor and natural historian, a naval professional, and a man of inherited wealth—started an extended self-experiment on diet and metabolism. Using an extremely simplified ration of just three food items, each of them undertook exercise while mostly fueled by a different part of the diet—one on a high-fat, one on a high-carbohydrate, and one on a high-protein ration. They performed heavy physical exercise and adjusted their rations according to their performance and their preferences. The results were conclusive: all three men converged on a very similar ratio of basic foodstuffs. Both the high-fat (the doctor) and the high-protein (the sailor) consumers were unable to finish their entire rations, while the high-carbohydrate subject (the gentry) felt hungry, suffered more from frostbite, and had a specific desire for fats. The subjects experimented with proportions, and eventually all three came to a ration whose ratios (by weight) were about 4:3:1 carbohydrate/protein/fat.

Although this may appear to have been a rather basic experiment, what to eat was a crucial question for expeditions right through the twentieth century. While the fundamental parts of the human diet had been identified and analyzed in nineteenth-century laboratories, by 1911 the detailed processes of metabolism remained a mystery, the exact roles of fats, carbohydrates, and proteins were unclear, and micronutrients were unknown. Food was a crucial and complicated technological issue for expeditions, not merely as a fuel for extreme physical exertions, but also as a vital psychological prop: boring, unpalatable, or unpopular food could cause serious issues with morale, while unfamiliar or contaminated food could cause morbidity and even mortality—dysentery, after all, caused the death of oxygen pioneer Alexander Kellas in 1921. So, while simple in design, this 1911 experiment was also rigorous: the men were together for twenty-four hours a day, there was no possibility of cheating by eating extra food or lying about the amount of exercise done, and weather conditions (the exercise was done outdoors) were very carefully noted.

And yet this study was never published in a journal about nutrition, or even as a note in the *British Medical Journal* or *Journal of Physiology*. If you want to find out about it, you have to read not a scientific paper, but a biographical account of the study, which was published in 1922, over a decade after the experiment ended. This failure to publish may be in part because nine months after this food study, two of the participants were dead and the third traumatized by his experience, particularly after recovering the bodies of his friends. The experimenters were Apsley Cherry-Garrard, Henry Bowers, and Edward Wilson, members of Robert Falcon Scott's ill-fated *Terra Nova* expedition (see fig. 5). The foodstuffs they had tested were "Antarctic" biscuit (135 pounds), pemmican (110 pounds), and butter (21 pounds), plus salt and tea.[1]

Scott had asked this winter party to experiment with sledging rations on their extraordinary expedition to collect Emperor penguin eggs from the birds' breeding grounds in Cape Crozier—the trip Cherry-Garrard later called "the worst journey in the world." Their results informed the ration

5. *Left:* Wilson, Bowers, and Cherry-Garrard before setting out on the "worst journey in the world." *Right:* The same team on their return, eating something other than pemmican, biscuits, and butter. (Both photographs by Herbert G. Ponting, 1911. © Scott Polar Research Institute, Cambridge, UK.)

5. (*continued*)

packs taken four months later on the attempt at the South Pole, on which Bowers and Wilson died with Scott, in a tent just eleven miles from a food and supply depot that had been refreshed a few days earlier by Cherry-Garrard. It would be easy to point to the disastrous and tragic outcome of this expedition as the reason why the results of its dietary (and other) studies were never formally published, but such absence in the published record is by no means unique to nutritional studies (Henry Raymond Guly has calculated that a third of the bacteriological studies carried out in Antarctica in the "heroic age" were never published).[2] As chapter 3 showed, such information is largely exchanged through word-of-mouth, informal networks, biographies, conference papers, and so on. This chapter, too, is concerned with how knowledge travels, but in this case the focus is on the translation of such knowledge from one location to another—from application in the Antarctic to a ration pack for Everest; from local, and sometimes indigenous, knowledge to placeless, globalized science.

Technologies of survival, including food, are reinvented as they move from place to place. Chapter 3 described the links between Antarctic and Himalayan expeditions in the twentieth century—often the same people were involved, similar research programs were pursued, and work done in one area was frequently ported into other environments. Consider, for example, the way INPHEXAN acted as a springboard for Silver Hut, or the use of Scott's Antarctic kit list by the 1920s Everest expeditions as a starting point for their own plans. (Antarctica was also a site for altitude research: the first medical studies at the US Plateau Station, one of the smallest and most isolated bases in the 1960s, were on high-altitude pulmonary edema.)[3] Likewise, the expertise gained in one area was usually considered portable to the other: in the 1920s the meteorologist George Simpson explicitly suggested that his Antarctic experience of using scientific instruments in adverse conditions would be of use to the planners of the early Everest expeditions.[4]

As we have seen, the Alps were used as a staging point for research in the high Himalaya, so we might expect the more accessible parts of the Arctic to have been used as a site of experiment and practice for journeys to the Far South. Indeed, the Arctic and sub-Arctic were used in this way, so the relative absence of the Arctic in this story so far may be puzzling. The reason, in part, is that the Arctic is a very different sociopolitical space from the Antarctic: vastly more heterogeneous, as, at the time of this writing, eight countries have territory within the Arctic Circle (if we include Iceland's tiny slice of Grimsey Island), and that number has waxed and waned with global political events and conflicts. In terms of terrain, the North Pole is the inverse of the South— approached first over land and afterward over water and ice. Crucially, too, the Arctic is populated: given the importance of acclimatization theories to Western physiologists working in the Antarctic, the fact that indigenous peoples made the northern circumpolar region home has had consequences for extreme physiology work as well as technologies of survival.

Of course, there are also similarities between the Arctic and Antarctic (and between polar regions and high altitude)—the major physiological challenges remain, after all, the cold and the physical and medical requirements of long-distance travel in isolated places. But it is the differences between the Far North and Far South that actually made the Arctic such a prime spot for the practice and testing of exploration technologies and techniques. Its greater accessibility, using either one's own national territories or those of friendly nations, meant that equipment and team members for expeditions elsewhere were frequently tested in Norway, Greenland, or Alaska. Meanwhile, the knowledge and expertise of indigenous peoples could be studied and exploited for the purposes of territorial claims at the other end of the earth, a practice, I suggest, that we should consider alongside genetics and pharmacology when we think about the phenomenon of bioprospecting.

Not all of these survival technologies were the output of or were directly connected to extreme physiology research. But they were technologies often analyzed by the scientists discussed in this book (or by colleagues who were materials scientists or nutritionists), as we shall see specifically in the case of food and clothing; or they were day-to-day technologies used by extreme physiologists whose material reality affected the work they were able to do in the field, as we shall see specifically in the case of shelter. These technologies were also confounding factors in physiological experiments: inappropriate rations brought expeditions to a premature halt, while, in an ironic twist, physiologists began to worry that the effectiveness of cold-weather clothing actually prevented White bodies from acclimatizing to Arctic or Antarctic climates.

One technology acts as an excellent case study to provide an overview of the ideas in this chapter, and it is an item from the 1911 winter party's experimental ration pack: pemmican. An indigenous technology, appropriated from the North, it was later reinvented, exported to Antarctica and high altitude, and turned into an industrial product of economic and even military significance. Indeed, when in the early 1850s a radical UK medical journal, the *Lancet*, sponsored a series of exposés about the quality of common foodstuffs, among the routine domestic substances such as coffee and bread that it discussed was pemmican.[5] Pemmican was of interest to the *Lancet* because the Royal Navy used it in its standard rations, so any adulteration had military as well as medical consequences. In its most basic form, pemmican consists of dried animal protein that is ground or pounded into small flakes or a powder, mixed with molten animal fat, and then allowed to cool and harden. Variations often involve added carbohydrates, such as pea flour or dried fruit, and the pemmican itself is usually served as a base for a soup or a thicker "hash" or "hoosh." Europeans first encountered pemmican in the seventeenth century, as it was a travel food commonly used by various indigenous peoples in the regions that are now in the north of the US and Canada. While many cultures developed similar ways to preserve and transport meat, it was North American pemmican that was "bioprospected" by European colonial powers. Comparatively lightweight and economical, given that it is extremely calorie dense and high in protein, it was used widely by European trappers and hunters, particularly in Canada. Consequently, it became an important store item for the big fur trading companies: by the first decade of the nineteenth century, the North West Company was using almost 40,000 to 60,000 pounds of pemmican every year to fuel its canoeists, trappers, and hunters; its rival, the Hudson's Bay Company, was consuming nearly 100,000 pounds of pemmican by 1840.[6]

Pemmican rapidly became an economically crucial foodstuff, and its production and distribution was the motivation for armed conflicts between

European colonials in North America as well as between Europeans and local peoples (indeed, the conflict between the local Métis and English-backed
colonials between 1814 and 1816 is sometimes referred to as the Pemmican
War).[7] When it began to be used as a food for sea voyages as well as land expeditions, it took on direct military significance, particularly in the Royal Navy—
hence the *Lancet* analysis. In Europe, in the middle third of the nineteenth
century, pemmican was absorbed into new laboratory-based investigations of
metabolism and nutrition and attempts to make a new, efficient, condensed,
modern foodstuff.[8] Many would-be food entrepreneurs and chemists attempted
variations of pemmican or other "meat biscuits" or tried to condense the
"goodness" of animal flesh into liquid tonics—Liebig's Extract of Meat being
the classic, later to become Oxo.[9]

   While mid- to late nineteenth-century explorers' handbooks offer advice
on making one's own pemmican in wild environments (the first edition of
Francis Galton's *The Art of Travel*, published in 1855, contains detailed descriptions of pemmican production apparently gathered from a Hudson's Bay
Company employee),[10] by the time European explorers started making serious
attempts on the North Pole, the South Pole, and the "third" pole of Everest,
the pemmican they carried was commercially made, came prepackaged from
major food firms such as Cadbury's or Bovril, and was often provided free in
return for advertising opportunities. With the discovery that pemmican is an
inappropriate foodstuff for high-altitude and hot-weather expeditions—for
biomedical reasons that will be outlined below—it became strictly an Arctic
and Antarctic ration, and in the twentieth century it was rapidly replaced by
other forms of condensed and dried food.

   The American explorer Robert Peary reported feeding the same pemmican
to dogs and men in the early twentieth century—noting his dissatisfaction
with some commercial American suppliers, one of which sent him pemmican
containing broken glass, identified rapidly by the human consumers and suspected as the cause of death for several dogs. (Sharing pemmican with dogs
was a constant source of risk: because it was a bland food, the explorers salted
it heavily, which was a cause of death among "Eskimo dogs" with no previous
experience of salt.)[11] Within a few decades this shared dining practice had completely changed: by the mid-twentieth century, if explorers took pemmican
at all, they took it *only* as food for sled animals—although on rare occasions,
their human companions reported eating "dog pemmican" in extremis. By this
point, though, pemmican had been reinvented—no longer the super-efficient,
scientific, mass-produced modern food of exploration, it was instead a marker of
hardship, suffering, and endurance. "Old school" explorers would be celebrated
for a preference for pemmican, and "new school" explorers would be mocked,

as their desire for more varied or more palatable foods was used by veterans of the poles to demonstrate their lack of robustness (and, by implication, masculinity). In a survey of the FIDS team in the late 1940s, even Vivian Fuchs reported that "no one found pemmican really palatable, and all were hungry to some extent."[12] A few years later, the South Georgia Survey (1953–54) reported that some men went hungry rather than eat pemmican during a ration trial. In the previous survey of 1951–52, it had been noted that the leader (explorer— and Dick Barton voice actor—Duncan Carse) was significantly older than most of the participants and had (largely stoical) attitudes toward food that they did not share. Clearly irritated by the experience in both years, Carse commented that "no major modification of a basic diet should be perpetrated in the field without first calculating and having some regard for the calorific values and vitamin figures involved" and that "a torrid passion for pickled onions is no justification for their supplanting pemmican."[13]

This story arc, from indigenous knowledge to reappropriated imperial science to traditional marker of a lost, robust, masculine past, can be found in many technologies of survival, and it describes, I argue, a form of bioprospecting. Historians and sociologists have so far tended to use the term *bioprospecting* narrowly to refer to seeking substances related directly to pharmacology or genetics; initially invented to refer to the hunt for medically or economically useful chemicals or genes in non-Western environments and populations at the end of the twentieth century, the term has been broadened by historians to encompass the early modern European attempts to exploit botanical resources for economically or medically useful substances.[14] But the story of pemmican is a clue that expanding this term to include other forms of indigenous knowledge—including technology and cultural practices—broadens our understanding of the micro- and macro-circulation of scientific knowledge, objects, and theories. At times, as we shall see throughout this chapter, the "local" origin of survival technology was an important component of its perceived value—"authenticity" sometimes serving as a proxy for "effectiveness."

As was demonstrated in the previous chapter, scientific expeditions to extreme environments were spaces in which "local" experience and expertise were highly valued and conserved. On occasion, expertise was also allocated to non-Western participants, especially in the process of moving technologies from the Far North to the unpeopled continent in the Far South. Of course, this allocation of technical expertise came with caveats: there are traces of "noble savage" imagery in discussions of Inuit or Sherpa knowledge that still figure indigenous knowledge as instinctive or natural rather than scientific and rational.[15] As in the case of, say, a cancer cure extracted from an Amazonian flower, indigenous knowledge could be made into a Western discovery by passing it through analytical laboratories and industrial factories, ending up as a commercialized and branded product whose origins are obscured—like "Cadbury's Pemmican." As with pharmacogenetic bioprospecting, the appropriation of survival technologies could also be deeply exploitative and could carry significant risks to people and ecosystems. Western mountaineers have occasionally looked at the lists of Sherpa dead—their deaths caused by the fact that climbers need to use the very bodies of local people as an essential prop to their expeditions—and wondered if this use is an ethical practice. And as survival technologies have made sites such as the South Pole or Everest more accessible to more people, they have had a concurrent environmental impact. Even pemmican tells this story: the demand for it was, although not its sole cause, a major contributing factor to the extinction of Canadian bison herds, again a pattern seen in the most controversial forms of pharmaceutical bioprospecting, which threaten ecosystems for the sake of commercial drug enterprises.[16]

The local knowledge that explorers sought out can be roughly divided into three categories: embodied knowledge, environmental knowledge, and survival technologies. *Embodied knowledge* here implies *biological* adaptations to environments, including racial characteristics or ethnic adaptations, and is discussed in detail in the following chapter. *Environmental knowledge* should be read to cover specific local knowledge—including information that is geographical (locations of water sources, safe routes and passages) or meteorological (signs for fair or poor weather, and so on). Finally, *survival technologies* can be material objects (snowshoes, sleds, clothing, kayaks, pemmican) or practices (care for clothing, dog handling, what time of day to start a hike). While beneficial racial adaptations remained a controversial topic in evolutionary discussions, it was relatively easy for Westerners to accept that indigenous populations might have specific environmental knowledge. For example, early Himalayan explorers frequently mention their use of pilgrims' trails or shepherds' paths, assuming that local people—perhaps over generations—would have found the safest or shortest routes over mountainous areas.

This chapter considers the hunt for and reinvention of local and indigenous environmental knowledge and survival technologies, and it therefore fits the Arctic more firmly into our evolving story of extreme physiology and scientific expeditions. In particular, it demonstrates how understandings of the local were created. Some of these definitions were quite new; for example, in Antarctica, where there was no indigenous population, "local" knowledge and experience could include that gathered in the Arctic by Western explorers or refined over generations by Inuit peoples. This chapter also takes seriously the notion that the concept of bioprospecting can be used to help us understand how practices move around the globe and, in particular, how they can be reinvented to enforce various identities, just as pemmican has been indigenous, traditional, modern, *and* a marker of robust White masculinity.

Hierarchies of Bodies, Hierarchies of Knowledge

As outlined in chapter 1, by the late nineteenth century, theories of human acclimatization generally assumed that the White body could adopt practices and technologies that would allow survival in non-temperate environments, but that these environments were inherently unhealthy, leading eventually to the sickness and deterioration of the White body as well as to the racial inferiority of the non-White indigenous populations. Much of this discussion focused on heat and humidity; mid-altitude environments appeared more often as a boon, as when White populations fled to hill stations in India to avoid the endemic hot-weather diseases, and there was almost no discussion of the long-term effects of cold climates.[17] Acclimatization (and later, tropical medicine) was primarily a colonial enterprise, so without serious European colonial interests in the Arctic Circle, Western biomedicine concentrated on survival in Africa, South America, and Australasia (although work on the equivalents of "colonial medicine" in the US—on acclimatization and survival—included proportionally more cold-weather work due to Alaskan and Arctic influences).[18] Broader Western interest in the survival technologies and local knowledge of indigenous peoples of high-altitude and cold regions therefore increased only in the closing decades of the nineteenth century as European and North American travelers began, in larger numbers, to explore the Himalaya and the Antarctic as well as the Arctic Circle.

The very process of mapping the high Himalaya demonstrates the ways in which local knowledge was used and ranked. The Great Trigonometric Survey, an extraordinarily ambitious program to map British India, was carried out in many instances by British-trained Indians. It produced the first estimates of the height of Everest, and while the theodolite measurements of the mountain were made in late 1849 and early 1850 by a British surveyor, Mr. J. O. Nicholson, the calculations were conducted by the "chief computer" of the survey, Calcutta-born mathematician Radhanath Sikhdar. It was his mathematical work that identified Everest (then known as "Peak XV") as the highest mountain in the world, effectively creating the goal that Western climbers would seek out, unsuccessfully, for more than a century.[19] But while Western explorers were willing to accept that local peoples had geographical and environmental knowledge, it proved harder for them to accept that these peoples would have invented new technologies or generated technological advances. The survey was a highly technical enterprise, and while historians have reclaimed the roles of Indians in redesigning, adapting, and improving the instruments and methods used, their White contemporaries were less generous in assessing local peoples' abilities to contribute to Western science.[20] While many were flattering about the work ethic, fast learning, bravery, and improvisation of trained Pundits, it was clear that this appreciation was framed within an understanding that measuring, mapping, and mathematics were inherently *Western* sciences imported to a new space.[21]

Local technologies and local peoples had their White champions. Dr. Tom Longstaff, one of the first writers in English on the problem of mountain sickness and doctor to the 1922 British Mount Everest expedition (and son of one of the major funders of Scott's *Discovery* expedition, Llewellyn Wood Longstaff), used a report on one of the first Western climbing expeditions to the Garhwal not just to praise, but to rank the abilities of, local peoples. His expedition included several members of the Fifth Ghurkha Rifles, who

> never failed us, they never complained, and they never lost their cheerfulness. Without them we could have done very little. They were superior to the best Garhwalis I have met, and even to the Bhotias, so I need hardly add that they bore no resemblance whatever to the Kumaoni or the down-country native.[22]

Further, Longstaff noted at least twice that the routes his climbing team planned or took turned out to be paths already discovered by local shepherds.[23] (Environmental knowledge was not the sole privilege of local *men*: the British Arctic explorer Sir Edmund Parry reported that while his boat was trapped in the ice in the 1820s, one "Eskimo" woman, who "was a very intelligent draughtsman," drew an accurate representation of the territory later named Melville Peninsula.)[24]

Longstaff's climbing team also included three Alpine guides (Alexis and Henri Brocherel, of Courmayeur, and Moritz Inderbinen from Zermatt). Extreme exploration often disrupts traditional senses of *local* knowledge with an expectation that specific geographical and environmental knowledge can be transported from one location and reinstalled in another. It was assumed that (White, European, largely working-class) Alpine guides would be useful colleagues to have on expeditions thousands of miles away from the Alps— and in more unusual cases, the reverse was also assumed, as Longstaff considered that Subhadar Karbir Burathoki of the Fifth Ghurkha Rifles "counted as a guide" because of his experience climbing not just in Kashmir and Karakoram, but also with Sir Martin Conway in the Alps.[25] While in many cases guiding knowledge did seem to transfer from one continent to another, by the second decade of the twentieth century Europeans were sufficiently impressed by the climbing abilities of local people to begin to think that importing Alpine guides to the Himalaya was not cost effective.[26] Dr. Kellas—the first champion of oxygen for mountaineering—began to recognize and publicize the superior climbing abilities of the Sherpa people shortly before his untimely death in 1921 on the British Mount Everest reconnaissance expedition.[27] (Kellas also recommended choosing Buddhist "coolies" over Hindus, as the latter were "handicapped by a comparatively rigid diet which in some respects is unsuitable for high altitudes . . . as it is extremely difficult to cook the nitrogenous vegetable foodstuffs above 16,000 feet.")[28]

Yet even as Sherpa porters effectively became a standard "technology" incorporated into European (and later American) expeditions to the high peaks of the Himalaya, their contributions were still conceptualized as a form of *embodied* knowledge—commentators pointed out that the Sherpa explored only at the behest of Westerners and thus were "natives who usually knew as little as [Western climbers] about the area and could therefore hardly be recognized as 'guides.'"[29] Through the mid-twentieth century Sherpa engagement with technology was at best mentioned as an inconvenience—when explorers cited the difficulty and expense of ordering high-altitude boots for porters, for example, or adjusting Caucasian-model face masks to the "flat noses of the Sherpa"[30]; or as an encumbrance—for example, in the many stories of Sherpa struggling with oxygen equipment, "meddling" with technology in transit, or breaking and stealing items.[31]

A somewhat different situation emerges when we look instead at high-latitude rather than high-altitude populations. While the people usually referred to in Western scientific publications as "Eskimo"[32] were frequently represented as uncivilized, primitive, and "simple," there was a much broader recognition of their environmental expertise and advanced survival technologies. This appreciation was, of course, moderated by racial and xenophobic assumptions about indigenous peoples, which could lead to a strange duality of attitude among Western travelers. Other historians have commented on explorers' use of Inuit technologies, particularly their extremely visible use of fur costumes, which were complicated forms of display (discussed further below) marking the explorers' rugged masculinity, but also "mask[ing] native contributions to polar exploration even as [explorers] literally wore the evidence of these contributions on their bodies."[33] Doublethink was endemic in Western reactions to "Eskimo" cultures: as briefly mentioned in the previous chapter, the Inuit-reported evidence of cannibalism on the Franklin expedition was dismissed by many, Charles Dickens famously included, as a fantasy by "savages." And yet at virtually the same time, the American Arctic explorer Charles Hall was able to craft an apparently more believable fantasy that Franklin may have survived, rescued by the "Eskimo" and taught to survive by these "iron sons of the north."[34]

Although the US gained Arctic territory officially with the purchase of Alaska from the Russians in 1867, North American explorers had been sailing to the Arctic throughout the nineteenth century—many in search of Franklin, and many with explicit scientific goals (at least until these fell out of fashion at the turn of the century). Encounters with Eskimo were an expected part of travel and adventure narratives, particularly from the mid-nineteenth century, when, as the exploration historian Michael Robinson points out, attitudes toward the potential "extinction" of the American Indian population meant that immigrant Americans were thirsty for stories of indigenous peoples.[35] Hall, who famously spent time "living with the Esquimaux" (the title of his 1864 book was *Life among the Esquimaux*), used his experience with Inuit survival technologies as a crucial part of his argument that Franklin could still be alive; indeed, Hall's failure to raise funds for his search expedition meant that he took no crew along and was almost entirely reliant on indigenous people.[36] Their kind treatment of this visitor had an unpleasant sequel, as Hall persuaded a Nugumiut family to return with him to the US, where he displayed them as an addendum to his talks and public lectures. Their journey eventually resulted in the death of the mother, Tookoolito, from an infection she picked up in America. (It is worth pointing out that this family was a long way from being the "simple" natives Hall suggested they were, having already traveled to the UK with a British captain some years before; and they were not only fluent in English, but had also met Queen Victoria and Prince Albert.)

Despite benefiting enormously from the knowledge of indigenous peoples, Hall played down their role in his various successes, not least because stories of survival and daring are more dramatic if the protagonists are not helped every step of the way by friendly local experts. A later generation of explorers, epitomized by Robert Peary and his rival Francis Cook, were more explicit about the debt they owed to local technology, particularly when it came to clothing and transport. While both still told stories about indigenous peoples that served their needs—for example, contrasting the luxuries of civilization against the harshness of "savage life" to emphasize their own rugged manliness and exceptionality—both also freely admitted the superiority of Eskimo-made fur clothing and dog-sledding.[37] Many a White explorer used a technique, highlighted by the historian Janice Cavell, of "demonstrating the superior qualities of his [own] race" by "combining civilised intellect and foresight with aboriginal survival strategies."[38] In 1956, Raymond Priestley, speaking in his capacity as president of the British Science Association, claimed that a "remarkable generation of Arctic explorers arose in Cambridge" in the 1920s who "evolved an efficient technique of living on the land, and on the sea in kayaks; they *out-Eskimood the Eskimos* [my emphasis]."[39]

While Western explorers had a tendency to "rank" the usefulness of indigenous populations at altitude (for example, Longstaff's opinion, above, that Ghurkhas were better than Bhotias, who were, in turn, better than Garhwalis), Arctic peoples often remained more homogenous in reports, whether in the *Lancet*, the *Polar Record*, or the many popular books about Arctic exploration and "life with the Esquimaux." And, of course, reports in European languages were almost exclusively written by Westerners *about* indigenous peoples and their technologies, rather than *by* indigenous peoples. The one notable exception is perhaps Vilhjalmur Stefansson, a North American explorer born in Canada to immigrant Icelandic parents (who had named him William Stephenson—he took up a more Icelandic name in college). Stefansson conducted extensive ethnographic work among indigenous Arctic populations and, as indicated by his name change, was able to claim a closer common heritage with circumpolar people than most contemporary North American explorers. But Stefansson clearly presented his work as anthropological observations, not autobiography, and so it formed part of the translation of Inuit technology and survival techniques into Western practices—that is, the "combination of civilised intellect" with indigenous knowledge that made using such technology acceptable.[40] Stefansson promoted the "native Eskimo" sledge as better than the North European "Nansen" style for rough ice—but only "as modified by Peary."[41] Likewise, while the "native" "is familiar with local conditions and not afraid of them," his environmental knowledge is still limited:

> Perhaps the chief superiority of a white hunter over a native is in ability to find his way about. By timing himself carefully, by making either a written or mental record of all distances and angles of travel, the white man will frequently know just what direction to turn for home when the native is completely muddled.[42]

Indeed, one analysis of the British Arctic expeditions between 1919 and 1939 concluded that "the most successful expeditions achieved their objectives without being dependent . . . on the use of Eskimo (Inuit) as guides," partly because an overreliance on the Eskimo "tended to concentrate the party's efforts on travel and hunting rather than on survey [*sic*] or other scientific work; there are no bears or seals on the hill tops or rock outcrops far inland."[43]

Whatever their apparent limitations, indigenous Arctic technologies and local knowledge remained of interest in the decades prior to the turn of the twentieth century, although relevant mostly to a still fairly small cadre of explorers in the Far North and to a curious reading public. These technologies of travel over snow and survival in cold climates became more valuable forms of local knowledge around 1900, with the revival of interest in the South Pole, and from around 1920, with the refocused attention on high-altitude expeditions. Of most interest, then, was not the *environmental* knowledge of the Eskimo, but their *technology*, whether applied to transport, shelter, or nutrition.

Reinventing Local Knowledge

While responses to indigenous expertise obviously varied—some explorers, and perhaps slightly fewer scientists, appreciated and celebrated local ingenuity—indigenous survival technologies generally experienced one of three fates: erasure, reinvention, or being "proved right" by Western science. These patterns are by no means limited to exploration science: there is a long-standing trope in Western thinking that represents "traditional" knowledge, especially in colonial settings, as static, so that it can be contrasted with the apparently natural progression of Western science. In this setting, local knowledge, while valuable, is either a remnant of extinct ancient superior cultures or a stagnant tradition.[44] This position can be difficult to sustain in relation to exploration technologies, especially in the case of local technologies that need to be moved from one place to another. In these cases, even Western participants are acutely aware of regional and temporal variations—and will invest heavily in them, debating issues such as which breed of sled dog is the best. Likewise, survival technologies, and their discussion in biographies, journal articles, or conversations at the Explorers Club, inevitably involve alterations to suit the conditions of the field, often while an expedition is still in progress. The strong military connection to extreme physiology also means that sometimes the results of extreme physiology research were kept secret for long periods, or were shared only with allies in wartime—even after the war there was a risk, or so physiologists thought, that important findings might be "buried" in military reports.[45]

It was therefore easy to erase the indigenous origins of survival technologies, or represent them as a basic concept that had been "improved" by the application of Western scientific principles.[46] There are reflections here of a similar process in "discovery" narratives, especially in the natural and geographical sciences: while settlers and colonials are well aware that indigenous peoples "discovered" animals, plants, or geographical features long before Westerners, a distinction is drawn between "scientific" or "rational" local knowledge and that of the indigene. Alternatively, indigenous knowledge could be "proved right" by the application of Western science, as we saw in the previous chapter with the location of Franklin's ships *Erebus* and *Terror*. The failure to take indigenous knowledge seriously until it has been, like pemmican, refined through Western science has sometimes proved dangerous: in the 1830s the American explorer Elisha Kane set out to disprove the "superstitious" Eskimo taboo against eating polar bear liver, only to find himself extremely sick as a result.[47]

Survival technologies are also, by necessity, *hyper*-local—they may be altered on the spot to deal with a particular microclimate or an unusual resource (such as a dead polar bear). While some anthropologists and ethnologists have described in detail the specific local technologies of particular indigenous groups, there is a tendency in Western literature to gloss or homogenize local practices, even though sleds, foods, shelters, and clothing vary not only from group to group (including expedition to expedition), but sometimes from day to day. So, for example, a survey of *komatik* technology (the "traditional Eskimo sledge") published in the *Polar Record* as recently as 1998 claims that the *komatiks* being built in Resolute, NWT, were "in general design and construction" identical to those built in the eighteenth century; yet less than a page later the same author says that "the Eskimos made use of whatever materials were at hand," describing sledges made from a whale jawbone, from "frozen fish rolled in sealskin or walrus hide," and with "sledge shoes built from reindeer antlers rather than mud."[48] It is hard not to imagine that a jawbone and a tube of frozen salmon might be different enough as travel technologies to create sleds adapted for specific sorts of terrain, yet they are subsumed, again, into a gutted understanding of the richness of indigenous technology that favors the "general design" and not localized ingenuity.[49]

A closer look at the technologies of shelter illustrates the paths that indigenous technologies can take when adopted by Western scientists and explorers. While the word *igloo*, as one author noted in a 1938 article in the *Polar Record*, "is one of the few Eskimo words that most white people understand," it is nonetheless a word that has held very different meanings in different circumstances.[50] By the turn of the twentieth century, Western writers with an interest in snow houses were bemoaning the fact that "the details of construction . . . and the more minute control of climatic conditions, have not, as yet, received adequate attention"—attention, of course, from Western science, as presumably Inuit peoples had by this point in history spent quite some time considering both these factors.[51] Explorers were keen to rationalize and explain the technology of the snow hut: in 1939 a French mountaineer, Louis Malavielle, published an article titled "Vacances en igloo sur le Mont-Blanc" in the main periodical of the French Alpine Club, which, as the title suggests, outlined his and his wife's experiences of building and inhabiting snow houses in Alpine sites above 3,200 m. (Again, it is notable how a woman's contribution is erased through the convention of publishing: this is clearly a joint venture, yet the man is the sole author of the article.)[52]

Translated at length in the *Polar Record*, Malavielle's article provides extremely detailed guidelines on the building of such shelters—from the size of the teeth on the snow knife used to cut blocks to the choice of snow, where, "contrary to Eskimo opinion, Malavielle regards all snow as usable, except that which is powdery and that which has thick layers of ice."[53] This article is one of many popular and semipopular representations of the "igloo" that ignores the origins of the word (*iglu* being a far more generic word, meaning "dwelling," while snow houses have their own specific terminology) and cements it in the Western imagination as a specific sort of dwelling, usually made from ice blocks and conventionally dome shaped. (Unless otherwise indicated, from this point onward my use of the Westernized spelling either indicates such a domed snow house or is the word used in the original text.)

Despite Malavielle's insistence that "the technique [is] essential for mountaineers as well as for polar travellers," there is little evidence of igloo building on high-altitude mountaineering expeditions through the twentieth century, partly for logistical reasons.[54] Explorers needed shelter during the largely snow-free trek to the mountains as well as during their time on the slopes where snow might be available. Certainly the building of snow houses is never mentioned among the preparations for the British Everest expeditions and related reconnaissance trips, although a great deal of time was spent researching and testing tent technologies. While proponents insisted that a skilled Eskimo could erect a functional igloo within thirty to forty-five minutes, tired European mountaineers and Sherpa unfamiliar with the technology clearly found tents a more reliable and efficient form of shelter. European travelers did sometimes build snow houses for amusement, as during downtime on the Swiss 1952 expedition to Everest, when André Roch "busied himself with constructing an igloo."[55]

Snow houses of various kinds were relatively more common as auxiliary buildings and sites for scientific work than for sleep and socializing. As early as 1893 Nansen's Arctic expeditionary team from the *Fram* built "a snow house . . . on the [ice] floe for magnetic observations."[56] Snow houses also appear in Antarctica, again, nearly always as accessory buildings not for residential use. For example, on the Australasian Antarctic Expedition (1911–14), Douglas Mawson reports that an "Igloo" was built by Alexander L. Kennedy, Charles T. Harrisson, and Sydney E. Jones to function as a magnetic observatory. Its construction took the three men the best part of five days, rather suggesting that Malavielle's estimate of thirty to forty-five minutes was somewhat overoptimistic.[57] Later in the expedition, another igloo functioned as a temporary shelter for the men sinking a geological shaft.[58] Elsewhere on the continent, Felix König, a mountaineer and member of the Second German Antarctic Expedition (1911–12), built an igloo to live in when the expedition ship *Deutschland* was trapped in ice—although this was read as part of a persecution complex and a desire to distance himself from two warring factions on the expedition.[59] More collaboratively, in the late 1950s William Siri attempted to build an igloo during the INPHEXAN expedition, while on the polar plateau in Victoria Land, "in anticipation of extremely bad weather . . . in the event our tents were blown away."[60] (Despite the availability of detailed instructions, construction again seemed not to be as simple as authors like Malavielle claimed, and Siri was "glad the igloo was in the middle of the Antarctic continent where no one could see it.")[61] Siri's contemporary on the ice Desmond "Roy" Homard, who was part of the TAE crossing, made a better attempt using a saw to cut ice blocks—Homard was also an engineer, which may have given him an advantage over the physiologist Siri.[62] Much later in the century members of the fractious IBEA occasionally built "igloo toilets" when they stopped long enough in one location, although these were attempts at privacy rather than fully functional and weatherproof construction.

These scattered references make it clear that by the second half of the twentieth century Western explorers were using the term *igloo* to refer to very basic ice caves and other temporary dugout shelters rather than the more sophisticated domed, ice-block structure described by Malavielle, which is the clichéd image of the igloo in Western culture. There is an irony in the nonspecific meaning of this borrowed word, in that it actually better reflects the original word *iglu*, but has been made vague by accident rather than by a better understanding of Inuit culture and linguistics. In his 1940 version of the *Arctic Manual*, written for the US military, Vilhjalmur Stefansson included a section bemoaning this misuse of the term *igloo*: "There is a doubly unfortunate practice in our books and speech to use the Eskimo word *Iglu* (Igloo) for a special type of house which we suppose to be or think of as being peculiar to the Eskimo."[63] The first problem this creates is that "we are then making a narrow, specific use of one of the broadest terms possible" (e.g., *iglu*, which indicates "dwelling" or "house"), and the second is that "a writer will call an iglu that type of Eskimo house with which he is most familiar"; that is, writers will fail to understand either the broadness of the term or the sheer heterogeneity of housing technology across the circumpolar region.[64] Even in the building of snow houses, however, Stefansson finds space for Western superiority—pointing out that such constructions are traditional in only certain parts of the Arctic, he suggests that "an American Boy Scout is more likely to have ideas on how to go about it than the average Eskimo boy of Alaska, Siberia, or Greenland."[65] (It should be noted, however, that he thought one reason for this was the "civilization" of circumpolar people, which led to the loss of indigenous cultures: "The eastern or western Eskimo, if he has heard of snow houses at all, thinks of them as unfashionable and hence inferior.")[66] Of course, references to the "igloo" on a Western expedition are sometimes not references to any sort of snow house, but rather to the brand and style names of various commercial tents. That tents should have remained more popular than igloos as expedition shelter is not surprising—not least because their major advantage is that they can be tested, and their erection practiced, in temperate zones before travel. Clearly, the experience of Westerners with snow house construction showed that it was not a simple practice, but a highly skilled form of building that required, at the very least, careful choice of location and availability of appropriate snow.

The slippage of the meaning of *igloo* from carefully created dome house to basic ice dugout to mass-produced tent is an illustration of the ways in which indigenous technologies can be stripped of their complexity as they are adapted or appropriated by Western science. Frequently they can be simplified, as in the case of the shift from a semipermanent, carefully designed snow house to a dugout, in which, although the simplification is necessitated by Western skill deficiency, the effect is to create what is actually a very basic form of the original technology while retaining the original name. This, of course, has happened twice for the igloo, first with the loss of the original diversity of the term *iglu* as it became *igloo*, and second with the bastardization of the word *igloo* into dugout, cave, or shelter. Other forms of flattening or loss occur when indigenous technology is combined with Western science, as in the cases of both pemmican and dome tents: while some "igloo" tents were inspired by the strength and construction principles of domed snow houses of various kinds, the effect is sometimes to erase that inspiration and use the original technology more as a totem—effectively trading on the "noble savage" image of indigenous peoples while making sure that the technological knowledge remains identified as White or Western.

The names of tents remind us that they are forms of local knowledge. For much of the first half of the twentieth century, British and North European explorers—both in Antarctica and at high altitude—tended to favor shelters known as "Whymper" or "Nansen" tents, named after the Alpine climber and the Arctic explorer, respectively.[67] In particular, Whymper tents, which were considered particularly good in terms of stability and wind resistance, were, according to contemporary gear makers, "not improved upon for mountain purposes . . . until the arrival of the geodesic dome in the late 1970s." (Nansen tents were more lightweight and extremely quick to erect, compromising the stability of the heavier Whymper style for these features.)[68]

The claim that these styles were not improved upon is not strictly accurate; in reality, every expedition tweaked and altered tent designs—not always to the benefit of the explorers—and reported back on changes that worked and changes that left them cold, exposed to the elements, or frustrated with the processes of erection. For example, the early British Everest expeditions in the 1920s introduced a tunnel-shaped entrance to the traditional Whymper, renaming the tent the "Meade," after the expedition's equipment officer, C. F. Meade (although the redesign was a group effort, particularly in response to George Mallory's specific complaints about the tent entrances).[69]

By 1955 the basement store of Everest equipment at the Royal Geographical Society held seven Meade tents, six in yellow or pink and one intended for photography, which presumably was of a darker color.[70] Tent color was a serious issue. In January 1953, just a few months before the British Mount Everest expedition was to set off, Pugh was still debating with the expedition leader about the color (and design) of his "physiological tent," arguing that "the green colour canvas does not permit sufficient lighting for gas analysis to be easily carried out throughout the daylight hours."[71] Even more subtle problems were discovered in the 1960s: during the Silver Hut expedition, rogue irregular sinusoidal waves were recorded on ECGs during some of the investigations of heart function at altitude. Although no reason was definitively discovered, the fact that these readings disappeared when the wind dropped led Jim Milledge to suggest that they were caused by static electricity generated by the largely nylon tent fabric.[72]

The ways in which experiential knowledge—zipper design, tent fabric, ration size—was shared between expeditions, including across national boundaries, has been outlined in the two previous chapters. Because experience could generate "local knowledge," explorers frequently preferred personal accounts to expert testimony; for example, in the preparations for the 1930s British Mount Everest expeditions, the transport officer Edward Shebbeare (deputy leader in 1933) spent time with the members of the German expeditions to Kangchenjunga led by Paul Bauer in 1929 and 1931. Shebbeare "took every opportunity of picking over their equipment & asking questions about it."[73] While he thought some of their equipment an improvement on the 1920s material the British used, when it came to their "Klepper" tents, they were "as good, but no better than [our tents] & the press-button fastenings are definitely worse than our tapes."[74] In these contexts, "local" is clearly a flexible category; just as survival knowledge from the Arctic was transferred to the Antarctic, there was significant exchange between Antarctic and high-altitude expeditions. When planning the 1935 British Mount Everest reconnaissance expedition, Frank Smythe (who was a member of all three expeditions in the

6. Andrew "Sandy" Irvine fixes an oxygen cylinder. (Photograph by Bentley
Beetham, 1924. © Royal Geographical Society, London, UK.)

1930s) insisted that "in particular we ought to study knowledge gained by
Arctic expeditions such as the British Graham Land Expedition. It was realised
last time how useful and invaluable were such things as Arctic tents."[75]

Again, though, such technology rarely traveled without being adapted. Virtually no form of technology was taken to the Arctic, Antarctic, or high altitude
and returned from there unaltered. This observation is core to the repeated insistence, which we encountered in chapter 2, that fieldwork is the only true test
of a technology or a survival technique. The ability to make on-site adaptations
or to use local materials in innovative ways is a skill admired in White and
Western explorers and scientists (even if it is erased for indigenous peoples).
Andrew "Sandy" Irvine, for example (see fig. 6), is credited with making essential changes to the oxygen equipment during the 1924 Everest expedition,
as "oxygen officer" Noel Odell wrote after his death:

> The writer would like to pay a tribute to the work done in reconstructing the
> apparatus by Mr Irvine, without whose mechanical faculty and manipulative
> skill, an efficient oxygen apparatus would hardly have been at the disposal of
> the expedition. Under difficult circumstances Mr Irvine has constructed an
> improved model about 5 lbs less in weight than the original.[76]

But not all ingenuity is created equal. In the 1950s Thundup, a Sherpa guide who worked with European teams in the Himalaya, made himself an outfit that the British climber Wilfred Noyce was fairly certain was fashioned from "the Cho Oyu sleeping-bags reported missing last year."[77] Even though down clothing was experimental technology for Western climbers at that time, Thundup's repurposing is represented as comic or criminal, not ingenious or inventive.

Clothes Make the Man; Clo Makes the Science

Clearly, clothing functions superbly as a marker of personal identity, although even its superior construction or ingenuity of design can be overridden by the ethnicity of the wearer. The adoption (or otherwise) of "local" clothing was an important technique used to maintain, alter, or assert national and racial identities throughout the decades of European colonial pursuits.[78] During the early history of colonial exploration, the functionality of clothing was also important in maintaining gender, as well as racial, markers in outdoor spaces: the pioneering "lady alpinists" of the late nineteenth century either managed their climbing in full skirts or carried them up the mountains in order to cover up their "indecent" climbing breeches upon their return home.[79]

While some technologies became rather homogenized across European and North American exploration teams (oxygen technology, for example, was consolidated into a few standard pieces), clothing, like food, remained an area of fierce dispute based on personal taste. British explorers remained skeptical, at least in contrast to the Americans and other North Europeans, when it came to fur clothing, preferring wool, cotton, and gabardine (particularly Jaeger-manufactured "performance" clothing). But elsewhere, bioprospecting was often the route to the design or purchase of the "best" clothing for either Arctic or Antarctic expeditions; explorers drew on the clothing practices of indigenous peoples, even if to later analyze and dismiss some outfits. While North Americans, for obvious geographical reasons, often cited the clothing of "Eskimo" in Canada, the clothing of other Arctic peoples—especially in Finland, Norway, and parts of Siberia—was also considered. Meanwhile, the organizers of the Japanese Antarctic Expedition of 1911–12 chose to take seal-skin boots made by the Ainu, an indigenous minority ethnic group.[80]

Discussions about clothing—whether in geographical and polar journals or in books about exploration—tended to the dogmatic, as individual explorers and their doctors, and accompanying physiologists, often fixed on one particular method of dress as objectively the "best." The desire here was to make what was unavoidably a subjective and embodied claim to superiority into something supported by objective evidence. This proved extremely challenging, not least because, just as had been the case for oxygen systems, the field and the laboratory rarely agreed neatly about the value of different technologies. Consequently, in the discussions about ideal clothing, there is often an appeal to multiple sources of authority to justify what is, perhaps, a personal preference (or a national tradition): the thermal properties of a particular fur or fabric might be cited alongside a story about its "traditional" use by "wise natives," backed up further by the author's own personal experience of his particular outfit. Stefansson is widely cited on the superiority of "Eskimo" clothing—his emphasis on the skill needed to maintain and care for furs meant that he could efficiently dismiss criticisms of fur clothing by claiming that if furs proved unsatisfactory, the problem lay with the inadequacy of an explorer's skills, not with the furs themselves.[81] These conclusions were accepted by many other writers of manuals and guides; for example, the American National Research Council's 1949 volume on the science of clothing assures the reader that the experience of explorers has shown "eskimo-style" clothes to be best in cold climates, but that "they require skilled making and skilled repair, as well as a certain wisdom in use, which has to be learned."[82] This argument is a variation on the "experimenter's regress": one successful experiment (not freezing to death) is taken to prove a theory (that fur is better), and any subsequent experiments that fail to reinforce or replicate this accepted result can be dismissed by claiming that the replicators did not have the right skill set, or the right technology, to properly conduct the investigation.[83] So, just as pemmican became the marker of a hardy masculine explorer, happy with deprivation, successful use of furs was sometimes configured as evidence of an experienced, careful traveler.

As was the case for *iglu*/igloo, the term *Eskimo* was flattened out in discussions of clothing; most Western writing on this subject homogenizes all the circumpolar peoples (or at least all the North American and Greenlandic peoples) as "Eskimo." In 1949, just a few years after the republication of Stefansson's *Antarctic Manual*, which is emphatic about the diversity of "Eskimo" cultures, the American scientist Paul Siple wrote, in a handbook put together for the US National Research Council, about the "Eskimo method" of dressing, by which he simply meant use of furs.[84] As well as being erased or reinvented, two of the three fates that I argue are usual for indigenous technology, fur clothing was also "proved right," but even this process could be used to tell stories about the superiority of Western bodies, if not Western technologies. In a 1966 study that was part of one of the Australian National Antarctic Research Expeditions (to Mawson Station, in this case), George Budd analyzed modern clothing and found it inferior to furs because the Australian team experienced a significant drop in skin temperature when they went outdoors. He contrasted this observation with claims that furs allowed "the Eskimo" to maintain an "'almost tropical' microclimate."[85] Instead of this research being read as an example of the superiority of non-Western technology, the implication drawn from it was rather that the increased experience of thermal stress meant that explorers in "modern clothing" were more likely to acclimatize to the cold than the thoroughly insulated "Eskimo" (as we saw in the previous chapter, the failure of experimental subjects to be properly cold stressed was an ongoing issue for studies of cold adaptation and acclimatization). What was gained in warmth from furs could be lost in biological adaptation, making the White visitor hardier than the "Eskimo."

The idea that an ideal set of clothing—perhaps somewhere between fur and gabardine—could be found using scientific analysis was rapidly belied by the experiential reality of exploring extreme environments. Just as those designing rations soon realized that food needed to satisfy emotional as well as nutritional needs, the feedback from clothing surveys made it very clear that personal preference played a significant role in explorers' satisfaction with their outfits. During preparations for the 1953 Everest expedition, Pugh and the rest of the organizing team took pains to survey explorers' opinions on food, equipment, and clothing. The diversity of their experience can be indicated by taking just one item of clothing as an example, in this case down jackets:

Six men want elastic wrist closures; two want both elastic and buttons. Seven out of nine men want zip instead of fly buttons. Three men suggest attachment of . . . hood to jacket by strong dome clips. Two men want more down filling. One wants nylon fabric.[86]

(We should note that the Sherpa were not surveyed for their opinions, although accounts such as those given above of their reuse of clothing "inherited" from expeditions suggests an active culture of innovation.)

This diversity of opinion did not mean that studies into the ideal clothing were abandoned, but rather that they were a repetition of the experience with oxygen systems, in which idealized (and simplified) laboratory studies struggled to mimic the real-world experiences of climbers, explorers, and mountaineers, but continued nonetheless. Clothing, too, was adjusted in the field when it did not suit particular bodies, tasks, or environments. Vivian Fuchs notes in his diary of the TAE in the late 1950s that his outfit had two bad

7. *Left:* According to the original caption, a "savage" child clad in dog skin, taken at Ponds Inlet, northern Baffin Island, Canada (1888). (Photograph by Walter Livingstone-Learmounth.) *Right:* British explorer F. J. Hooper in "modern" gabardine (1912). (Photograph by Frank Debenham. Both © Scott Polar Research Institute, Cambridge, UK.)

7. (*continued*)

pieces of design: first, his long johns did not button across the front, leaving
a gap, and second, he was using "the very stupidly made modern American
type of shirt which is split down the front."[87] This unlucky combination led
to him "almost [being] frostbitten in an unfortunate place and apart from it
being painful it took [him] sometime to thaw out with [his] hands."[88] Fuchs
records his intention to sew up all his shirts down the front. For similar reasons, even a technology as apparently simple as clothing also presents issues

of cross-compatibility between different systems. As the American explorer and scientist Paul Siple wrote evocatively in the 1940s:

> Unless the [clothing] assemblies are designed to match in function, great difficulty can ensue. Nothing is more disconcerting than to hurriedly manoeuvre through three layers, the windproof with a 2-inch vertical slit protected by a backdrop curtain of cloth, the wool trousers with a recalcitrant zipper, the top of which cannot be reached through the slit of the windproof, and finally stubbornly buttoned underwear.[89]

One challenge for physiologists and others trying to develop a form of "clothing science" was the need to find standardized metrics for warmth and comfort. In 1872 Francis Galton drew on experiments by Count Rumford (Sir Benjamin Thompson) on the thermal properties of fabrics for the final edition of *The Art of Travel*. These experiments, conducted in Munich between 1784 and 1788, involved placing a mercury thermometer in a bed made of various fabrics and skins, heating the bed in boiling water and subsequently placing it in ice, and recording the speed at which the temperature fell by a fixed amount of 135°F (for hare fur, this was 1,312 seconds, compared with the least insulating fabric, twisted silk, at 917 seconds).[90] This "thermal conductivity," measured in a variety of different ways, remained one of the few ways to talk about clothing scientifically until the development of the standardized "clo" unit by Gagge, Burton, and Bazett during the Second World War. This new unit was based on the met unit, where 1 met was "the metabolism of a subject resting in a sitting position under conditions of thermal comfort"; therefore, 1 clo was the amount of insulation needed for clothing to "maintain in comfort such a sitting-resting subject in a normally ventilated room."[91] In the early 1940s Burton also developed the copper mannequins that were widely used to test the overall clo value of outfits.

While it was largely laboratory-bound physiologists who did much of this initial standardization work, they had clear links to the network of exploration physiologists we have already encountered. At around the same time the clo was developed, Siple was in the Antarctic, based at Little America III, measuring heat loss and freezing rates in plastic tubes filled with water. His conclusions (whose publication was delayed for several years due to their relevance to the war effort) introduced the concept of windchill to thermal measurements—that is, the effect of air speed on the subjective and objective temperature experienced.[92] Siple's extensive studies on clothing, insulation, and thermal comfort in the Antarctic (and in cold-room laboratories) are widely cited, and the clo is widely used, yet both are also subject to criticism for being oversimplified representations of what increasingly revealed itself to be a complex reality. For example, in his TAE clothing study, Rogers points out that Siple's windchill factor makes no allowance for the effect of solar radiation, and he notes that during the TAE study, "the correlation between clothing worn and windchill was much inferior to the correlation between clothing and temperature, and this was true for all the data."[93]

Similarly, Pugh showed that altitude had an effect on clo estimations because the reduction in air density with altitude increased the insulation value of clothing (at Silver Hut, it would be about 17% higher than at sea level, according to estimations by Pugh's boss, Otto Edholm, and clo co-inventor Burton).[94] Even in the laboratory, estimating thermal comfort was extremely difficult: as we have seen in the previous chapter, Allan Rogers got different figures from different laboratories for the same clothing assembly when it was tested on the copper mannequin at the Institute of Aviation Medicine in Farnborough and at Wright-Patterson Air Force Base. He blamed this variation, of up to half a clo per assembly, on "the difficulties experienced in putting the clothing on the Farnborough mannikin, and, despite unpicking and re-stitching seams, there was probably a tighter fit . . . with a lower clo value as a result."[95] The challenges that Rogers faced in getting his data on thermal comfort and clothing analyzed (leading to that thirteen-year publication delay) is a salutary reminder of the complexity of these physiological and biomedical studies. Although field tests can at first glance appear "rough and ready" compared with finely controlled laboratory studies or complex mathematical predictions of the response of human skin to winds and low temperatures, their complexity is a consequence of researchers' attempts to model as closely as possible the lived experience of the explorer. Absent the controls of a laboratory, these studies sometimes use the very variability of the environment as a way of making their findings more robust and applicable.

As well as challenging the priority of the laboratory, these studies belie another cherished myth: the dominance of the gentleman amateur in European exploration. While there is clearly evidence of a strong nostalgic desire for explorer-heroes who face extreme environments while badly or inadequately equipped, the reality is that Anglophone and European teams of the twentieth century were almost universally supplied with the best technology of the day. Indeed, while the outfits seen in photographs of the teams from the 1920s and '30s can seem old-fashioned (even compared with the outfits of the 1950s expeditions), twenty-first-century research has literally recovered and metaphorically reclaimed these outfits. Samples of George Mallory's clothing were taken from his body during the 1999 Mallory and Irvine

Research Expedition (discussed in the previous chapter) and eventually deposited in the National Mountaineering Exhibition at Rheged in Cumbria.[96] Two of the few authors to have written extensively about mountaineering technology, Professor Mary Rose and businessman Mike Parsons, initiated the "Mallory Clothing Replica Project" to re-create this clothing and test its thermal efficiency, water repellency, and so on. Across laboratories (and climate chambers) at the Universities of Lancaster, Southampton, Leeds, and Derby, these re-created clothing assemblies were tested, and the research team concluded in 2004 that Mallory (and, by implication, Irvine) were adequately insulated against the weather conditions of Everest during the 1924 climbing season.[97] In fact, Mallory's outfit was considerably lighter than outfits made of modern fabrics, suggesting that the equipment of 1924 may even have had advantages over that of 2004—an opinion confirmed during the (by now apparently obligatory) field testing of the re-created outfit by British climber Graham Hoyland, who tried the gear around Base Camp and the Rongbuk Glacier, rather than at the summit, and was, in fact, too warm.[98]

Mythologies about choices, even those relating to the most basic and quotidian of technologies, can be used to tell socially or culturally important stories. The early British Everest teams were skeptical about the value of oxygen at least partly for quite rational, "scientific" and experience-based reasons, but that skepticism has also frequently been presented as evidence of "gentlemanly amateurism" and the class tensions and snobbery inherent in the elite climbing world. Likewise, differing choices of clothing have been mobilized to tell stories about national identity and manliness, but the tendency for early Antarctic British expeditions to avoid furs, while the Norwegians used them, seems to have had less to do with Nansen's appreciation of circumpolar folk wisdom (vs. British disdain for indigenous knowledge) than with the fact that the British were *man-hauling*, not *dog-sledding*, to the South Pole. Hauling sledges is a physically demanding activity, requiring a massive caloric intake, and in turn is a very effective source of body heat. Furs were too warm for the British teams when they took to dragging their sleds, but just warm enough for the Norwegians taking part in the comparatively sedentary activity of driving dog teams.[99] And, as every commentator has pointed out, often the "best" equipment is the best only if it is handled properly by explorers with adequate experience; often the most familiar technology is, in fact, the "best" one for an individual. On international expeditions, other sorts of compromises between familiarity and experience had to be found, as John Giaever, leader of the Norwegian-British-Swedish Antarctic Expedition (1949–52), explained:

The British had their ideas with regard to polar equipment and we had ours. And the Swedes, all the same, certainly had their ideas too. With polar equipment, as with so much else, the problems have more than one practical solution. Very soon, however, we agreed to a compromise, and choose the best after entirely objective testing, "the very hard way."[100]

It was the Swedes, on the Norwegian-British-Swedish Antarctic Expedition, who were responsible for sourcing the sledges, while the British organized the dogs (the Norwegians brought the boat). Even the choice of dog or man, or indeed, pony, as the motive force for sleds in polar regions is a good example of the imbrication of scientific analysis and national prejudice. Ponies were mostly a temporary feature in Antarctica; they had been used with some success in various Arctic ventures, and explorers like Shackleton and Scott favored native Siberian ponies, reasonably assuming that they would perform well in the cold of the Antarctic too—another example of the portability of local knowledge. While they proved useful in limited circumstances, the stories of the ponies are generally of injury, slaughter, and conversion into food for man or dog, and few explorers took them after the early 1910s. Otherwise, the history of transport in the polar regions has been one of man-hauling (or man-paddling, when it comes to the open waters of the Arctic), dog-sledding, and later, motor vehicles.

Scott's team has received much retrospective criticism for its choice of manpower rather than dog power, and both historians and contemporary commentators have associated the British preference as a result of—depending on your interpretation—squeamishness or sporting ethics. That it seemed brutally calculating (and "un-British") to use dogs when the plan was to drive them and then eat them was certainly a view expressed by early twentieth-century explorers (although Scott's views on dogs are, according to the historian Carl Murray, often quoted out of context; while he took ponies and dogs, and ended up man-hauling, Scott actually hoped to make better use of his *mechanized* transport).[101] Yet British explorers nonetheless did eat dogs and ponies, or at least used both as animal food on long expeditions. There was also an ongoing debate about whether the "Eskimo" treated their dogs well or brutally—authors' assumptions were based on a mixture of their own stance on Eskimo culture (advanced vs. primitive) and, of course, their limited experience of a diverse circumpolar population.[102] "Brutal" treatment was very much in the eye of the beholder, as British Arctic explorers, at least into the 1930s, would engage in activities such as breaking the back teeth of their huskies to stop them chewing on harnesses and tracer leads made out of sealskin.[103] Cruelty narratives extend beyond the

Arctic—Somervell's recollections of the 1922 British Mount Everest expedition included a donkey that collapsed, leading its driver

> in true Oriental fashion . . . to pull it up by the tail, causing the poor moke a certain amount of pain. I have always been particularly moved to indignation by cruelty to all sentient beings, and waxed very wroth with the driver, whom I regret to say I knocked down. He apologized humbly for not pulling the tail harder and more efficiently; the only explanation of my anger that occurred to him![104]

It is too easy to give a nationalist or "gentlemanly" gloss to decisions about equipment (or animals) that, on closer examination, turn out to be rather more complicated. At the turn of the twentieth century, it had not yet been established "scientifically" what the most efficient means of travel across variable ice and snow terrain was, and both dogs and motor engines required skilled and experienced users. The introduction of motor vehicles into Arctic and Antarctic travel was often a process of trial and error: just as with oxygen systems, clothing, or any other technology, "in the field" and often emergency adaptations had to be made to vehicles as the reality of extreme cold or difficult terrain proved too much for the theoretical speculations underlying the design of the machines. Just as with oxygen systems, a poorly chosen or badly designed piece of transport equipment—an aggressive dog or a malfunctioning tractor—could seriously hamper exploration and could even put lives at risk. The question of which mode of transport was "best" was a hyper-local decision, depending not only on the specific terrain and particular goal, but also on the inherent skills of the scientists and explorers on the expedition. Stefansson certainly applied a circular logic—an "explorer's regress"—in suggesting that furs were superior clothing and that anyone who disagreed simply did not know how to use them properly; yet it is demonstrably the case that individual variation in skill, preference, and experience had a significant effect—perhaps as significant as national tradition or gentlemanly ethics—on the choice and successful use of survival technologies.

Conclusion: Eat Local

In hindsight, and with the replication of the journey by Fiennes and Stroud, it has become clear that one of the reasons why Scott's polar dash team suffered so badly is that their rations were not sufficient to support the significantly increased caloric demands of man-hauling as compared with dog-sledding—despite the fact that Scott had organized on-site experimental research into nutrition: the ration experiment by Bowers, Wilson, and Cherry-Garrard that opened this chapter. Food is both a survival technology and a cultural product, as we saw above in the case of pemmican, which was comprehensively reinvented over the nineteenth and early twentieth centuries. (By 1949 there was even a vegetarian alternative, which was developed by Mapleton's Nut Food Company in Liverpool, advised by nutritionists from Cambridge University, and tested on glaciologists from Cambridge working in Jotunheimen in Norway in the early 1950s—who declared it "more appetizing" than their alternative, meaty Bovril pemmican.)[105]

Pemmican is also an example of apparently universalizable scientific knowledge (in this case nutritional analysis) that does not work in all locations. Despite being a core adventure food for the Far North and Far South, pemmican is not a good ration item for high-altitude work because the human sense of taste is altered by altitude (a challenge faced most often now by airline caterers). The loss of appetite was a known problem for early twentieth-century mountaineers. While polar explorers generally lost weight due to a simple calorie deficiency in their rations, mountaineers apparently lost weight because they chose not to eat the calories they had brought with them. Although in the 1960s the mechanism of this physiological shift was unclear, its reality was well known: as Pugh wrote, "All members of the wintering party" at Silver Hut "noticed impairment of appetite, particularly for fatty foods, and, as time went on, a marked preference developed for highly seasoned foods and condiments."[106]

Awareness of this problem was part of the reason why early twentieth-century Himalayan expeditions tended to take more varied—even luxurious—ration packs with them than did Arctic or Antarctic explorers. These choices are part of the mythology of choice that uses them as evidence of amateurism or gentlemanliness, or simply the quaint alienness of the past. Long lists of condiments, tinned luxuries, champagne, and foie gras are easily cited as humorous evidence of the different priorities of other times, but they are in fact often based on good scientific evidence about starvation, appetite, and altitude.[107] In the particular case of foie gras, in 1952 Pugh wrote to Miss M. W. Grant, a specialist in nutrition at the London School of Hygiene and Tropical Medicine, to ask about "haemoglobin promoting foods" and got the following response:

> The foods which are supposed to be of the most value when haemoglobin has to be made are (1) liver, kidney and gizzard, (2) egg, heart, and (3) apricots, peaches, prunes, apples. It looks as though you ought to have pate de foie gras

and caviare [*sic*] in the basic diet (more cheaply liver sausage and smoked cod roe)—Might be possible?[108]

Despite all the evidence, and all the concern, sometimes the wrong decisions were made about food. One of the many flaws of the IBEA in the late 1970s was that the team used French military rations.[109] These rations were canned, while most Antarctic rations were dried, and they proved to be difficult to cook and monotonous to eat in the polar region. Another source of resentment for the IBEA team members was that they were traveling with a French glaciological team who "used heated vehicles and trailers and ate food of excellent quality with wine and spirits to match."[110] One wonders why IBEA made such poor choices, psychologically speaking, but they may be another example of the significance of experience and informal knowledge—a great deal of food expertise was based on word of mouth and on surveying explorers to ask their preferences based on experience. Expeditionary archives often contain food questionnaires—similar to the clothing questionnaires—with useful entries suggesting, for example, that the best thing about Grape-Nuts is that "yaks like them."[111] While the INPHEXAN team was conducting its physiological tests, an otherwise unnoticed nutritional experiment was also happening in Antarctica, as the veteran explorer Sir Hubert Wilikins (then aged 69, and at one time a traveling companion of Stefansson) and two colleagues were camping at Cape Evans and testing a new set of dehydrated survival foods.[112] Wilikins's death the following year almost certainly prevented the publication of any information on these studies.

Beyond the need for appetizing and stimulating rations, by the 1910s there was still no consensus on even the basic principles of diet (as the research undertaken on the trek to collect Emperor penguin eggs at Cape Crozier demonstrates). Physiologists and nutritionists generally agreed that either fat or carbohydrates were the main source of motive power, but the role of protein was still unclear. Through the nineteenth century Western medicine had tended to advise low-protein diets for those traveling to hot and humid climates; this advice was justified by an appeal to various medical theories, but also satisfied the assumption that animal protein was a source of vital energy (and intelligence), as high red meat consumption was read by Western Europeans as a marker of civilization.[113] Travelers to India, sub-Saharan Africa, South America, and the Pacific were advised to consider a "less stimulating" diet of vegetable proteins as well as abstaining from alcohol. This advice, too, follows the pattern of bioprospecting I have laid out in the previous pages—initially it was based, at least in part, on the observation of local customs and survival behaviors.[114] Later research would suggest that a low-protein diet is in fact a good approach to rations for desert and high-temperature work: protein metabolism increases urine production, and so "water-protecting" rations for those in high-temperature, low-water situations (such as desert expeditions, or life rafts at sea) should contain high proportions of carbohydrate and fat and be low in protein.[115] This advice renders pemmican a poor choice for the hot-country traveler as well as the mountaineer; it is also another instance in which existing indigenous knowledge has been "proved right" only by passing through the laboratories and field studies of Western science.

The logical converse of the "low-protein diet for heat" hypothesis would be that high-protein diets are better for cold climates. This hypothesis seemed to be confirmed by the cultural practices of the peoples of the circumpolar region: while nineteenth-century explorers pointed to the use of lentils and pulses in tropical cultures, the "Eskimo" was often represented as consuming extremely high proportions of meat and animal fat (often in the form of blubber).[116] At the same time as Antarctic explorers were experimenting with various ration packs, physiologists examined indigenous Arctic peoples to try to assess the effects of their diet on their health, and specifically on their ability to survive in cold environments. The earliest comprehensive studies, and probably the most famous, were those by Norwegian doctor and physiologist Kåre Rodahl, who gained his doctorate with a thesis on the toxicity of polar bear liver—an infamous killer of explorers (especially those who ignored local advice).[117] Over the course of several years of living and working in Alaska, Rodahl and his wife, Jean, performed extensive physiological and medical examinations of indigenous peoples. One of the theories Rodahl proposed was that some local acclimatization to cold was a consequence of local high-protein diets. These diets seemed to correspond to a raised metabolic rate; that is, the body began to create more biological heat. The groups with the diets highest in protein (in this case, the people sampled in Anaktuvuk Pass) had the highest metabolic rates, whereas when indigenous peoples were put on traditional Western-style American diets, their metabolic rates fell to levels comparable to those of visiting Whites.[118] This important conclusion was part of a growing body of evidence that man can, and does, adapt physiologically and racially to heat and altitude, but technologically and culturally to cold—an idea explored more in the next chapter. As Rodahl put it:

> Our studies had indicated that the reason the Eskimo gets along better in the Arctic than we do is mainly because he is completely adjusted to the environment, and not because of any unique racial endowment.[119]

This line of research was continued, and pursued to an extreme, by Vilhjalmur Stefansson, who was converted early to the idea that the high-protein diet of the "Eskimo" was healthy and possibly superior to a fatty, sugary, Western diet. Stefansson vigorously fought against what he saw as a prejudice against pemmican as "unscientific" and "primitive" (this seems something of a straw man, given the widespread use of pemmican by explorers during this period). His campaign for the high-protein, high-fat diet was waged through publications, but also through the use of his own body, as in 1928–29 he and co-explorer Karsten Anderson lived for a year on a "meat-diet" that entirely excluded all vegetable products and most dairy products.[120] He helpfully translated the high-protein culture into practices acceptable to "civilized" North Americans by including a recommended five-course dining option for "hostesses" to copy, which finished with "*Dessert*: Gelatin, solely of meat origin and made according to a recipe which I got hold of somewhere."[121] Stefansson's appeal to traditional knowledge was combined with scientific studies and tests, but also based its logic on an assumption about the primitive ("stone-age") diet to which he assumed humans were hereditarily adapted; his work, and these assumptions are the origin of the modern Paleo diet.

White explorers chose particular survival technologies to bioprospect from the indigenous residents of cold and high and hot places. Likewise, this chapter has cherry-picked just a few examples—from tents to *komatik*—to illustrate the processes by which material objects, cultural practices, and even animals could be reinterpreted, reinvented, and moved around the globe for the purpose of scientific investigation and sporting exploration. Similar stories could be told about debates over ski design, hunting practices, and even the choice of llamas and yaks in marketplaces for treks to high altitude. A series of (sometimes contradictory) processes shaped and reshaped survival technologies. Some are processes we have seen at work in previous chapters: the repeated movement of technology—material, animal, cultural—in and out of laboratories and field sites as clothing is analyzed, tested, found wanting, redesigned, and retested in a pattern similar to that which we saw for oxygen systems in chapter 2. We have also seen the prioritization of local experiential knowledge—although in this chapter the identity of the local is complicated. Most paradoxically, local knowledge has become both globally portable and incredibly granular: technologies for survival in the Far North have been used, at least as starting points, for survival at literally the other end of the earth, as well as at high altitude. Yet at the same time, the local expertise of the dog-sledder when it comes to choosing clothing or rations has proved unhelpful to the man-hauler, even if he is traveling the exact same route. There are also clearly times when the local knowledge of the explorer and the indigenous population overlap. For example, the previous chapter discussed the practice of leaving food and other supplies for future, sometimes only imagined, travelers. Caching was not the prerogative of White or Western peoples in the polar regions or at altitude: one poignant counterexample was the attempts by two Sherpa to point Maurice Wilkins to a food cache in 1934. Wilkins planned to climb Everest solo, powered by mystical beliefs, and while the two Sherpa who accompanied him finally turned back and left him on the mountain, they first made sure that he knew the site of a "food dump" left by the British expedition of the previous year (he ignored this food, and died on the mountain).[122]

The appropriation of indigenous technology could simultaneously highlight its portability and its extreme localness. In a 1934 *Polar Record* article titled "The Eskimo Kayak," an anonymous author (probably the journal's editor, Frank Debenham, director of the Scott Polar Research Institute) emphasizes that "the kayak varies very much according to locality."[123] He then goes on to give a detailed account of the reconstruction of an "Eskimo kayak" by a Cambridge University student, who, having read an account of the "Eskimo roll" maneuver in a book about the British Antarctic Air Route expedition, spent much of Lent (Spring) term 1933 with friends trying to re-create the maneuver:

> They went further still, and taught themselves to execute the roll with the throwing stick instead of the paddle, and finally with the hand alone, a feat which only a very few of the Eskimo can manage.[124]

It was from these activities that the *Polar Record* author went on to provide an extended discussion of building and using an "Eskimo" kayak—in other words, from the embodied life experience of a young White Englishman, practicing in the River Cam, in England. Presumably these are the sorts of people Priestley had in mind in 1956 when he spoke of a Cambridge generation who "out-Eskimood the Eskimos."

So, at the same time as the category of "local knowledge" becomes more refined and specific, there is a tendency to "flatten"—to simplify and homogenize—*indigenous* knowledge. Some explorers and scientists, Rodahl being a solid example of both, laid great emphasis on the diversity of indigenous cultures and their micro-local adaptive technologies, through which Western commentators often made claims about their own experience and expertise. But just as often, the local variation of indigenous technologies was lost: "Maybe with the introduction of the motor toboggan we shall become like Rodahl's Eskimo," wrote J. R. Brotherhood, an environmental physiologist and colleague of Pugh at the MRC, in 1972, theorizing that the use of less physically demanding transport would make White Antarctic locals more likely to experience, and therefore adapt to, bodily cold.[125] "Rodahl's Eskimo" was, of course, a dozen different indigenous groups, and the work of both the Rodahls (Jean as well as Kåre) documented a diverse culture of adaptations, but the term remained useful shorthand nonetheless. Likewise, as we saw above, *igloo* became shorthand for any form of ice cave or shelter, while the domed snow house became its clichéd representation.[126]

Some Western technologies were also flattened. The Whymper, Meade, and Nansen tents seem, at first glance, to have been remarkably long-lived survival technologies. Yet they, like virtually all other survival technologies, from boots to rations, were tweaked for the individual preferences of explorers, the needs and priorities of expeditions (e.g., adaptations to ensure they were sufficiently ventilated or the right color for scientific work such as developing photographs), or the particular demands of the locality. The network of men (and later women) who engaged in this sort of work knew through experience and informal discussion that these adaptations took place, but it may have been less obvious to the broader public who consumed stories of exploration through autobiographies and expedition reports. While it's not clear why teams did not rebrand their equipment as, say, Meade 2.0, or give new tents entirely new names, there is clearly some desire—as we saw with pilgrimages to Silver Hut—to associate new expeditions with established heroes. For all the competitiveness of extreme exploration, and indeed the crucial importance of priority for scientific publications and discoveries, the appeal of nostalgia, the desire to claim and demonstrate a *pedigree*, is a recurrent theme in writing—formal and informal—on extreme physiology and expeditionary work, at least among Western scientists and explorers.

Associating oneself with previous explorers—whether by re-creating their routes, recovering their abandoned equipment, or simply using tents that bear their names—is, of course, a form of identity creation. Here, too, nostalgia plays an important role. Within the space of a generation from the first concerted efforts to reach the South Pole, those involved in exploring and exploiting Antarctica were already looking back to the "good old days" and the better personal qualities of previous generations. As we saw in the previous chapter, by the mid-1950s Fuchs was already missing a (possibly imagined) Antarctic population that was more rugged, more invested in the continent, that consisted only of "natural" explorers. Some of this hero worship also feeds narratives about amateurism, gentlemanly or otherwise: it is easier to criticize a modern generation for being lazy or cosseted in extreme conditions if one has a past generation as a comparator whose equipment, clothing, and technology can be represented as basic, unscientific, untested.

While any survival technology can be mobilized in this way (and most are), food is a particularly poignant example, and one that also draws in aspects of racial identity as well as class and gender. Pemmican again functions as an excellent example of this process through its reinvention in the twentieth century. By the 1950s, when Fuchs was bemoaning the disinterested new generation, pemmican had become a dog food, and its consumption by humans was a marker of a robust, manly previous generation willing to suffer hardships. This was not the end of pemmican's trajectory, however, as around the turn of the twenty-first century it experienced yet another reinvention, this time associated not with exploration, but with new dietary fads such as the Paleo diet. Allegedly inspired by "primitive" eating patterns and the work of Stefansson in the 1920s, these diets are high in fats and proteins and aim to reduce the consumption of complex carbohydrates: pemmican, a mix of animal fat and lean animal meat, has therefore become a foodstuff of health-conscious, wealthy urban populations.

Despite its extraordinary relocation, some of pemmican's core identities are clearly still important. As of 2017, the website of the Classic Jerky Company, maker of "Pemmican Beef Jerky," sells its vision thus:

> Not so long ago men spent their free time outdoors. They went camping, fishing and hunting. They breathed in clean air and felt that sense of calm that only comes from being in the wild. And wherever they went, they brought jerky. The snack you packed for every escape.[127]

Aside from the emphatically masculine tone, the branding is illustrated by a generic indigenous American chieftain in full-feathered headdress. As is so often the case, it uses indigenous peoples as a marker of authenticity, of closeness to a "natural" way of living, which can be a positive representation when taken in contrast to an effete, Western, technologized civilization. It uses an appropriated image of non-White people to demonstrate *primitiveness*—drawing on the ongoing appeal of the "noble savage."

This White civilized/non-White primitive dichotomy was not just maintained by camping equipment, adventure snacks, and exploration literature: it was often the fundamental assumption behind scientific investigations. Most obviously, it is visible in studies of indigenous populations during the International Biological Program. The IBP was a ten-year global project inspired by the IGY that sought to produce systematic knowledge about the complex ecosystems on earth.[128] The human part of this work came largely under the theme labeled "Human Adaptability," and while some studies did look at White and Western populations (particularly in relation to nutrition, growth pattern norms, and exercise physiology), there was a concerted effort to survey, sample, and systematize the indigenous populations of non-temperate areas. Time and again, the reason given for studying isolated indigenous populations was that they were closer to a shared human past; that to study Inuit or Australian Aboriginal or Fuegian was to perform a kind of archaeology—of culture and genetics—excavating a biological and sociological reality that was the "past" for White, Western man. As Joanna Radin has pointed out, these studies were given their urgency by the assumption that indigenous peoples were at immediate risk of extinction through assimilation.[129] While that meant the "dilution" of blood (more on that in the next chapter), it also meant the loss of traditional survival cultures; thus the IBP needed anthropologists as much as it needed geneticists to record the "disappearing" people of the hot, high, and cold areas on earth.[130]

This fear of extinction was a not a novel one inspired by the IBP; in the first half of the twentieth century, much writing about Inuit groups (and some other circumpolar peoples, such as the Sami) by Western observers expressed concerns about the loss of traditional skills.[131] In these writings, there is a noticeable difference in the ways in which Western explorers and scientists talked about the populations of high, cold, and hot climates. While populations in all three climates were at risk of losing their cultural distinctiveness and were vulnerable to the health effects of Western diets and Western sedentary habits, concern about the loss of useful *technologies*, rather than arts or cultural practices, is much more visible in writing about circumpolar peoples. Mountaineers did not express a concern about dressing Sherpa in "civilized" high-altitude boots in preference to their traditional footwear; anthropologists did not bemoan the loss of Aboriginal water cultures as a loss of adaptive knowledge for White settlers; but for an "Eskimo" to forget his sledging heritage was a serious issue. Perhaps it is overly cynical to suggest that the reason for this difference is that Eskimo technology proved such a useful resource for White exploration, but the distinction between cold/technology and hot/biology must have shaped at least some of these divergent reactions to cultural losses. The following chapter, with a focus on blood—literal and metaphorical—looks further at how such notions of the relationship between race and place shaped extreme physiology in the twentieth century.

FIVE

Blood on the Mountain

Toward the end of the nineteenth century, several European researchers began to produce evidence that the blood of people who had spent time at mid-altitude showed marked changes from its condition at sea level, most notably an increase in the concentration of red blood cells. In 1937 this change was still an intriguing physiological puzzle, and in an attempt to solve it, a German physiologist unwittingly saved his own life. Ulrich Luft (1910–91) was a young doctor who had qualified in 1935 and gained his Ph.D. in 1937 on the physiology of respiration (particularly anoxia); as a keen mountaineer, he got the opportunity of a lifetime when he secured a place on an expedition led by Karl Wein to Nanga Parbat.[1] Luft was invited by Hans Hartmann, the chief physiologist of the Luftfahrtmedizinisches Forschungsinstitute (Aviation Physiology Research Institute), and was tasked with a variety of physiological work, mostly studying the effect of altitude on changes in the circulatory and respiratory systems—a topic of keen interest to aviators as well as to climbers.[2]

The climbing team split up at Base Camp, and Luft remained at a lower altitude to work on the physiological research program. The seven other German climbers and nine Sherpa porters continued up the mountain, getting as far as the site they had planned for Camp IV (around 6,100 m above sea level), and began the process of laying the route to Camp V. Three days after the party split, Luft and "five fresh coolies with stores and the mailbag" continued up the mountain to rejoin the team.[3] At the site of Camp IV, they were met with nothing but a blank field of newly fallen snow. Luft dug down and began to find the tattered remains of tents; he sent a distress telegram to the German Alpine Society, which dispatched a "crack team" of climbers, led by the (in)famous mountaineer Paul Bauer.[4] Its effort was, in principle, a rescue mission, but in reality was an attempt to retrieve the bodies of the climbers—the entire team of sixteen had been wiped out in the space of a few minutes by a massive avalanche. Hartmann's broken wristwatch set the probable time of their deaths at just after midnight ("Hartmann's wrist-watch showed 12.20 when we took it from him, but in my pocket it worked again. The cold of the snow hard pressed round Hartmann's wrist had brought it to a stop").[5] Luft later wrote that they must have been killed instantly, without knowing what was happening, as "they lay peacefully in their tents, their faces showing no sign of fear of the approaching disaster."[6] The bodies of Adolf Göttner and Peter Müllritter could not be found, but the remaining five Germans—Hartmann, Wein, Martin Pfeffer, Günther Hepp, and Pert Fankhauser—were reburied together "in a tomb below an ice-block as big as a house," while Luft recovered some notebooks and scientific equipment.[7]

With his expertise in anoxia, respiration, and altitude, Luft suffered a bout of nominative determinism and went to work for the Luftwaffe during the Second World War. Immediately after the war, he was targeted for recruitment by the Americans as part of Operation Paperclip, a strategic program that identified scientists with expertise thought to be useful to the US, particularly its nuclear ambitions.[8] Luft's major contributions were in a rather different area of research, as his expertise on the physiology of extreme environments meant that he did a great deal of work for the American space program, including designing the physiological screening tests for potential astronauts. He seems to have fitted comfortably into the American physiological research scene and was clearly a popular teacher and mentor.[9] He is directly connected to the expeditions already discussed in this book, not only as a researcher who is cited by extreme physiologists, but on a more personal level, as he provided advice and physical testing for mountaineers on occasion (Tom Nevison, of the Silver Hut expedition, had had his body density measured by "Dr. Luft").[10]

This positive remembrance of Luft was made problematic at the end of the century, when newly released documents and ongoing investigations revealed the extent of, and criticized, US recruitment of scientists who had been involved in unethical Nazi war work. Various materials (including accounts by Luft's son, Frederick) made it clear that Luft had been aware of—if not an active participant in—the murderous researches into hypothermia and barometric decompression in concentration camps; evidence suggests that this was not just a matter of paperwork, but that Luft had actually watched footage of the low-pressure tests conducted in Dachau.[11]

The abandoned bodies of dead Sherpa porters, and the frozen and exploded bodies of murdered Jewish prisoners, are pointed reminders of the darker history of extreme physiology. While the ethics of this research are discussed further in the next chapter, Luft's story is an introduction to the themes that tie this chapter together. Here the focus is on blood, both literal and metaphorical. In its physical, material form, blood was a crucial part of extreme physiology studies; while we have already encountered it as a sample, a form of data, in this chapter it becomes a technology of survival. The study Luft was undertaking on Nanga Parbat as the rest of his team was crushed beneath an avalanche was a direct descendant of studies inspired by Paul Bert. Bert's research had itself been inspired (and partly funded) by Denis Jourdanet, who had developed a theory about mountain sickness that conceptualized it as a form of anemia—although an unusual one, as the blood of those at altitude tended to be thicker and darker than the blood of those at sea level, suggesting an increase in red blood cell production. Bert himself wrote that, based on his understanding of the homeostatic capacities of the human (and animal) body, such polycythemia (many blood cells) was a *theoretical* possibility at altitude. Bert encouraged the young French researcher Françoise-Gilbert Viault to test this theory, and Viault confirmed it by finding a "considerable augmentation" of the number of "globules rouge" in the blood of mid-altitude residents; indeed, his own red blood cell count increased up to 60% over the course of the weeks he spent at altitude in the Andes, reaching values close to those he found in local residents.[12] Several other researchers repeated this finding—including Nathan Zuntz—but by the turn of the century a question mark still hung over the mechanism of this polycythemia. Was it a genuine increase in the production of red blood cells, or was it rather a decrease in the other contents of the blood, particularly the plasma? Mountaineers, like anyone doing vigorous exercise, tend to be dehydrated by their exertions, which leads to a reduction in the liquid content of the blood, effectively increasing the proportion of red blood cells to plasma (this proportion is known as the hematocrit: high hematocrit values mean a high proportion of red blood cells to other parts of the blood).

In 1906 two more French researchers, Paul Caront and Clotilde Deflandre, outlined a theory that the mechanism of polycythemia is controlled by a hormone or similar signaling molecule, which they provisionally named *hémopoïétine*. Their theory meant that increased red blood cell production would be possible if the signaling molecule were up-regulated by oxygen stress, anoxia, or some similar environmental pressure. It was not until the middle of the twentieth century that erythropoietin (EPO) was established as this elusive signal for red blood cell production and its up-regulation was proved to be a genuine biological phenomenon. Once EPO had been extracted, purified, concentrated, and eventually artificially synthesized toward the end of the century, it became best known as a doping technology for elite athletes (although it also has uses in disease treatment). But before EPO doping became a practical possibility, physiologists and explorers considered other mechanisms for increasing their hematocrit in the hopes that it would improve sporting performance or speed up the process of acclimatization to altitude.

When Pugh returned from Everest in 1953, one of the first letters he responded to was a blood-related query from the Birmingham-based surgeon Dr. J. S. Horn. Horn, a founding member of the Institute of Accident Surgery in the UK, suggested that blood transfusions could be used to "pre-pack" the circulation of climbers heading to high altitude.[13] Pugh was skeptical, citing the challenges of this sort of intervention, but he did say that "the Germans" had tried it "many years ago," with limited success.[14] Although Pugh promised to follow up with "references" to the German research, in the end he could not locate them, which suggests that he may have been misremembering the work done by Luft and others in the 1930s. It certainly seems unlikely that the technology of transfusion was advanced enough in the 1930s for it to be attempted at high altitude (and a transfusion at sea level would "wear off" before a mountaineer in the Himalaya could benefit from it)—indeed, the struggles physiologists faced doing basic tasks such as taking blood samples through the end of the century indicate that a major transfusion would be a risky endeavor anywhere in the high Himalaya. But if blood doping never became a survival or enhancement technology for climbers, the exact opposite process—blood thinning—was tried in the 1970s. This chapter will explain how two completely opposed medical interventions, concentration and dilution, could both be logical responses to the challenges of extreme environments.

Meanwhile, in its symbolic form, "blood" is a proxy for, and sometimes a literal biomedical marker of, race and ethnicity. As this chapter will show, physiologists from all over the world asserted that the blood of people at and from altitude is different from that of sea-level populations, and this measure of "difference" was used to make assertions and claims about the relative worth of different ethnic groups as well as sweeping claims about the history of human evolution. Blood research was significantly shaped by racial science, and in at least one case presumptions about the homogeneity of indigenous peoples have directly affected the research progress of Western physiology. While this chapter recognizes exploration science as a deeply racialized practice, blood has a final meaning, in the form of *menstrual* blood. Menstruation was an excuse to exclude the "unstable" and "disruptive" bodies of women from expeditions and experiments, but as this chapter will show, women were widely present in both experiments and expeditions, although they were often doubly invisible as women and people of color. The following sections outline the seesaw history of blood as an assistive technology and follow through to see how race and gender intersected in the work of extreme physiologists.

Blood, More or Less

One dramatic feature of twentieth-century biomedicine that is rarely visible in historical works is the sudden, extensive interest it showed in blood. From the early searches for traces of bacterial or malarial infection in smears of blood, through the discovery of immunological blood types and their relationship to heredity, to the extraction and sequencing of DNA from white blood cells at the end of the century, billions of blood samples have been taken from populations across the globe in the hope of defeating disease or uncovering the evolutionary history of humankind. Explorers were not exempt from this new vampirism, and most of those mentioned so far in this book had their blood taken at least once. What makes the studies considered here slightly different from the broader global story of this new fascination with blood is that extreme physiology research focused more often on hormones, blood gases, and sugars than on hereditary markers. The study of heredity was occasionally important, and for blood studies on expeditions it was usually played out on the largest scale possible, as racial and evolutionary studies predominated. But the strong focus of the history of biomedicine on molecules and genes through the end of the twentieth century should not distract us from the fact that blood itself, as a whole, macroscopic object, was turned into an assistive technology just like oxygen. And like oxygen, it provoked questions about what constituted "cheating" and what was "natural" in terms of physical performances and embodied skills.

By the early twentieth century, it was clear that one of the first responses of a human body newly relocated to altitude was polycythemia, although the mechanism for this change had not yet been established—was it upregulation of red blood cell production or a process of dehydration? The letter from Horn to Pugh suggests that physiologists (and maybe mountaineers) had begun to consider the possibility that transfusions could be used to "pack" blood and give climbers a head start on the process of acclimatization. But, as indicated above, the technology of transfusions lagged far behind the theoretical exploration of their potential. It was not until the 1940s that the two key requirements for successful widespread use of transfusions—rapid blood typing to prevent adverse reactions and effective means of blood preservation for transport and storage—were adequately developed.[15] Although blood typing was not important for autotransfusions (transfusing individuals with their own blood), storage still mattered, and it was not sufficiently developed to make self–blood doping on a mountainside a realistic proposal, at least for the first half of the century. So, while EPO use and transfusions attracted the attention of those involved in competitive sports, there was relatively little interest in blood manipulation in Western climbing or expeditionary science through the middle of the twentieth century.

Much more attention was paid, as earlier chapters have demonstrated, to oxygen and respiration than to hemoglobin and circulation—at least in terms of producing usable technologies or practices to aid mountaineering and exploration. Indeed, the mixed results of polycythemia studies (described in more detail below) meant that many climbing scientists had a rather skeptical attitude toward the potential value of transfusions by midcentury, when they started to become a technical possibility. In 1963, in response to a query from a doctor about blood packing, Tom Hornbein wrote that "people have been hunting for some such panacea for many years," but "none have proved to be successful," and, furthermore, that

> very few studies have been performed to determine how significant polycythaemia is to altitude adaptation. . . . The problems of red blood transfusion, of course, are not to be ignored, even beyond the question that they may not be exceedingly helpful. Dangers of transfusion reactions, hepatitis, and the fact that increase in hematocrit due to transfusion tend [sic] to suppress the body's own red blood cell producing activity suggest that it might not be terribly beneficial.[16]

Hornbein's response indicates a core problem for would-be transfusionist climbers: homeostasis, again. The body has many compensatory mechanisms, and consequently, when the balance of a system is disrupted, it frequently has knock-on, and sometimes quite unexpected, consequences in other regulatory systems. In the situation Hornbein cites, the natural production of red blood cells would be down-regulated by a high hematocrit, so the transfusion of blood would have no overall midterm effect, as the body would simply stop producing red blood cells until equilibrium was reestablished. While a transfusion's effects might last long enough for a sprint event, they would disperse before a mountaineer could get to even mid-altitude.

Indeed, the next significant attempt to manipulate blood for the purpose of climbing occurred only in the late 1970s and involved not blood packing, but the complete opposite: hemo*dilution*. Once again it was German researchers who were credited, at least in the Anglophone world, as the first experimenters in this area. In 1980 the American Medical Research Expedition (AMREE) applied for a grant from the American Lung Association for funds to study hemodilution, citing then-unpublished German research into the topic as its motivation. During the German studies, climbers had a fixed amount of their own blood removed and were then infused with blood *plasma*, which dropped their hematocrit from 59% to around 51%.[17] One of these "heme-reduced" climbers apparently set a new height record for climbing without oxygen.[18] Unofficial correspondence reveals that the reason the AMREE organizers were interested in this work was that

> the Germans have been fooling with this, but they have done no really scientific, controlled studies. The Germans state that hemodilution is great—makes you feel like a million and enables you to climb like the wind.[19]

It might seem counterintuitive that both blood packing and hemodilution could improve climbing performance, but this paradox is a consequence of the fact that the immediate response of a sea-level-acclimatized body to altitude—an increase in red blood cell production—is not without negative side effects. Two homeostatic mechanisms compensate for the low P_{O_2} at altitude: one changes the strength of the bond between hemoglobin and oxygen, enabling red blood cells to "grab" more oxygen at the lungs; the other, which we have discussed here, up-regulates the production of red blood cells, "packing the blood."[20] But both these changes can have negative consequences, which particularly affect the extremities of the human body. The tighter bond between hemoglobin and oxygen means that sometimes, in the capillaries and extremities of the body, red blood cells do not give up their oxygen to the tissue efficiently, starving some areas of oxygen (this effect is eventually canceled out as the oxygen affinity of the hemoglobin shifts in the opposite direction after a period of time in hypoxic conditions). "Packed" blood is literally denser, physically thicker, and more viscous, and these properties can retard its circulation in the smallest vessels—again affecting the microcirculation of the body and causing oxygen starvation in the extremities. While this phenomenon can obviously have multiple negative effects on the body, there is a double effect on the extremities in cold environments, where low circulation and low oxygenation can exacerbate dangers such as frostbite. Consequently, frostbite—or the lack of it—was a key measure of success for blood manipulation on mountains. The write-up of the 1975 Austro-German expedition to Kanchenjunga, which tested the thinning of blood by both infusion of plasma and "dilution" using water, claims that "none of our infused or diluted climbers suffered from frostbite even though the members of these two groups spent altogether 1.3 man-years at altitudes between 5,500 and 8,500 m, in temperatures as low as [minus] 36°C."[21] (This expedition was a relatively successful one that put nine members of its eleven-person team on the summit.)

The AMREE expedition, a few years later, conducted blood studies at Everest Base Camp, around 5,400 m above sea level, at roughly the same height as those on Kanchenjunga, and showed that above a certain point (50% hematocrit), polycythemia no longer provided a performance advantage.[22] AMREE's conclusions on the reverse manipulation, hemodilution, were more muted than the "feel like the wind" conclusions of the Austro-German expedition, in part because of the problems AMREE encountered in its attempt to study multifactorial homeostatic mechanisms. Just as in the nineteenth-century studies of polycythemia, the issue of dehydration was crucial. The routine dehydration of climbers means that their blood can become concentrated, so was hemodilution actually providing a novel benefit to the climber, or was it merely a process of counteracting dehydration? As the Kanchenjunga study put it, "To 'autodilute' by increasing intake of fluids by mouth seems the easiest method of hemodilution."[23] In other words, the rather complicated (and sometimes still dangerous) process of transfusing serum into climbers could be mimicked simply by having them drink more water. By the 1980s the results of all this work on hemodilution were evident only in concrete advice to climbers to avoid dehydration by sufficient oral intake of liquids. But elite mountaineers had already adopted that practice—one reason for the success of the Austrian Mount Everest expedition in 1978 (including the first successful climb to the summit without oxygen) was that the team paid particular attention to the threat of dehydration.[24]

Neither concentrating nor diluting the blood proved to be an uncomplicated or unambiguous assistive technology on the mountain for Western explorers or scientists. The very homeostatic principles that had suggested transfusions as a mechanism for assisting climbs in the first place also mitigated their eventual effects in the real world. And at the same time the practical applications of these ideas were failing in field tests, the very theory behind them was also being questioned. Until the middle of the twentieth century, the phenomenon of altitude polycythemia was an accepted fact; repeated studies, from Viault's onward, had shown that long-term residents at altitude, including indigenous mid-altitude residents, had high hematocrit readings, and while the exact measurements varied among individuals, increased hematocrits were a universal finding in sea-level residents at altitude. What remained controversial, then, was whether there was a significant difference in responsive potential between people born at sea level and long-term or indigenous residents at altitude. While climbers' hematocrits often reached values close to those found in permanent mid-altitude populations, there was still room to argue that certain people had a hereditary advantage over others, and that some thresholds of response were effectively genetic or evolutionary; that is, that sea-level indigenous peoples could acclimatize so far, and no further. On the other hand, the similarity of body responses and hematocrit values among populations also gave plenty of ammunition to those who wanted to argue that the performance of sea-level populations, particularly White Europeans, would eventually match that of indigenous populations and that total acclimatization to altitude was possible.

The reason this debate lasted in this form for so long—effectively a century, from the middle of the nineteenth to the middle of the twentieth—was that Western scientists focused exclusively on the bodies of South American indigenous peoples, using them as models for all populations at altitude. With one exception—a short study by the British doctor Thomas Somervell, published in 1925 and apparently roundly ignored—all blood work on mid-altitude residents was done, until the 1950s, on Andean residents.[25] Sherpa blood, when it was eventually tested, would prove a revelation. When Pugh wrote to J. S. Horn in 1953 to explain why transfusions would be no good to climbers, one of his justifications was that he had performed some of the first (and at this point still unpublished) hematocrit studies of Sherpa on Cho Oyu in 1952 and had found that although Sherpa gave extraordinary performances at altitude, their hematocrits remained at about the level expected in healthy sea-level populations.[26] While it took the rest of the century for physiologists to find an explanation for this surprising finding, by the 1960s and '70s researchers were wondering if polycythemia might be a *mal*adaptation to altitude—hence, of course, the interest in hemodilution.[27] It is clear that the use of the Andean body, rather than the Himalayan body, in physiological studies significantly shifted the focus and conclusions of circulatory physiology in the first hundred years of polycythemia studies, but it also had consequences for the study of race and evolution.

Acclimatizing to Racial Science

The homeostatic principle may be a surprising core for racialized beliefs about the superiority or inferiority of ethnic groups, but that nineteenth-century principle did help shape—or at least reinforce—presumptions about the hierarchy of the human races. As outlined in previous chapters, a combination of colonial experience, new evolutionary theories, and experimental physiology led to a tighter association between race and place; in essence, the result was a form of environmental determinism, the assumption that different groups of people are biologically adapted to the ecosystems and environments in which they, or their ancestors, lived. Western science asserted the superiority of the White, European, temperate-zone dweller and consequently took that body as the "normal" human form. This is where homeostatic representations mattered: it was the sea-level hematocrit that was assumed to be the "normal" and the mid-altitude reading that was the deviation or adaptation. With that assumption in place, it is easier to conceptualize indigenous mid-altitude populations (or populations in tropical or extremely cold environments) as somehow compromised, as their physiology adapts to a non-ideal environment.

The twentieth century saw shifts in scientific thinking about race. One optimistic interpretation of these changes is that of Nancy Stepan, who suggests that the idea of "fixed," definable races fell out of favor after the Second World War and was replaced by a "new science of human difference," which involved a more fluid, genetically defined understanding of mutable "populations" with ever-fluctuating "gene frequencies" that allowed diversity and unstable boundaries between groups.[28] While this new understanding was present in extreme physiology work and publications, there is also evidence that older, more fixed attitudes, particularly environmental determinism itself, remained; even those researchers who denied the reality of objective racial groups could still agree that individual genetic makeup would be shaped by environmental pressures. Deeply racialized—and overtly racist—views are easy to find among explorers, who frequently talked about Sherpa, "Eskimo," and other indigenous peoples in terms that were both patronizing and essentializing. Assumed, or real, adaptations to extreme environments were also interpreted in ways that reinforced racial hierarchies.[29]

The first and most obvious example, which comes from high altitude, is one we have already encountered: Joseph Barcroft's assertion in 1925 that "all dwellers at high altitudes are persons of impaired physical and mental powers."[30] This claim, as we saw earlier, acted as a red rag to Carlos Monge Medrano, who responded by organizing a series of expeditions and studies to specifically counter what he saw as a generalized (racial) slur against the high-altitude populations of the Andes, particularly Peru. We can possibly give more of the benefit of the doubt to Barcroft than Monge M. did—Barcroft's statement is grounded in his view of homeostasis as a zero-sum game, as epitomized by the changes in blood discussed above, in which an increase in red blood cells can be a benefit at altitude, but can also have negative consequences in terms of circulation. His claim that dwellers at high altitude are "impaired" could therefore be read as a comparative statement: they are "impaired" compared with how *they* could perform in an oxygen-rich atmosphere. It is not (necessarily) intended to imply that dwellers at altitude are impaired in comparison with dwellers at sea level, or to be read as a broader statement about the overall racial "ranking" of sea-level or altitude dwellers. However, Barcroft's statement is also reminiscent of a long-standing conflict between European scientific ideas and South American investigators over the question of high-altitude "degeneration."

Degeneration was the outcome that many in the nineteenth century saw as the inevitable result of prolonged residence away from the temperate zones of Europe and North America: a nebulous process, it usually indicated some sort of decline, decay, reversion to more primitive ways of being, congenital sickness, mental retardation, or sterility. And while White bodies were at risk of degeneration, indigenous peoples were already degenerate—generations of living in non-temperate climates had an inevitable, and hereditary, effect. Tropical degeneration gathered the most scientific (and consequently, historical) attention, but some European scientists also wrote about the possibility of degeneration at altitude—that populations exposed over the long term to "rarefied air" would become weakened, as individuals and as a race (this belief, however, was always in tension with the persistent use of mid-altitude spaces, particularly in the Alps and in India, as "healthy" alternatives to crowded and polluted cities, or as pure spaces and spas for the treatment of respiratory diseases).[31] One of the earliest concerted efforts to counter the assumption that rarefied air led to degeneration was that undertaken by Dr. Daniel Vergara Lope Escobar in Mexico from around 1890 through the early twentieth century. Escobar studied adaptation to altitude, including polycythemia, read the works of European scientists such as Bert, Mosso, Viault, and Jourdanet, and took extensive anthropometric and physiological measurements of mid-altitude Mexican populations to defend them against accusations of inferiority.[32] Vergara Lope concluded that there was a "law of compensation"—a form of homeostatic rebalancing—at work, so that the disadvantages of living at mid-altitude were compensated for by permanent or temporary changes to metabolism and physiology (a conclusion rather in advance of, but in general agreement with, Barcroft's own).

Given that Monge M. was part of a local movement—which stretched back to the nineteenth century—to defend the mountain and plateau dweller, it is perhaps ironic that the first result of Monge M.'s reading of

8. "Conquer Everest," a 1976 board game by Capri (*left*). While the game emphasizes tents and oxygen as technology, it also contains many instances of "superstitious" Sherpa who steal equipment or are "lost" along with equipment (*right*)—the next event on the route up the mountain is "Fall off Ice-Ladder. Lose 2 Sherpas with 2 Rope Gear." (© the author 2017.)

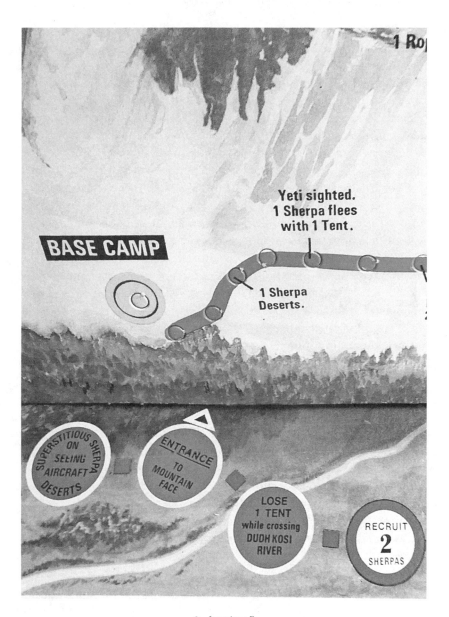

8. (*continued*)

Barcroft was that he "discovered" and named a specific high-altitude disease: chronic altitude sickness, or Monge's disease.[33] Monge M. also reformulated the evidence on adaptation to altitude in a broad, sweeping, environmentally deterministic explanation for the history of South America. In 1948 he published *Acclimatisation in the Andes: Historical Confirmations of "Climatic Aggression" in the Development of Andean Man*, with *climatic aggression* being a term Monge M. uses to talk about the influence of the high-altitude environment on short- and long-term physiological processes—in other words, the way the human body responds to the "attack" or "aggression" of the environment in which it finds itself. Monge M. makes a clear argument for a specific racial type, which he calls Andean Man: "being different from sea-level man, his biological personality must be measured with a scale distinct from that applied to the men of the lower valleys and plains."[34] This difference is caused by the "deterministic forces of an oxygen-deficient environment; the climatic imperatives determine and sustain an adaptive equilibrium." Monge M. goes on to make a specific dig at Barcroft, saying, "Ignorance of these postulates has led eminent men of science to make shocking errors of interpretation."[35] Climatic aggression itself works to create an equilibrium of human ecosystems—it creates a challenging environment for sea-level populations to survive in, but at the same time Andean Man, or populations fully adapted to altitude, also experience "climatic aggression" when they move from the mountains down to sea level. As man has an instinctive desire, Monge M. claims, to match his constitution to the "air temper" of a region, Andean Man's relocation to sea level can be only temporary. Repeatedly, however, Monge M. makes a case for the physical superiority of Andean Man in any environment—adapted to a hypoxic situation, he (and we should acknowledge here that there is no reference made to Andean *woman*) is a "natural athlete" because athletic training is merely the process of "inuring [athletes] to anoxia."[36]

Acclimatisation in the Andes is a complex work, not simply because of its blend of science and history, but also because it serves a political purpose—it is a mechanism for Monge M. to critique recent historical and contemporary policies (or lack of them) relating to the movement of labor, the practices of public health and hygiene, and the acknowledgment of indigenous alongside immigrant traditions. His work is not necessarily representative of Western attitudes toward environmental determinism or adaptation to altitude in the late 1940s; the foreword to the book, by Isiah Bowman, a geographer and president of Johns Hopkins University (whose press published *Acclimatisation in the Andes*), is somewhat defensive about its content. Bowman acknowledges that Monge M.'s book will "reignite" debates about determinism, and that it could seem old-fashioned to Western readers in a context where "the word 'influence' or the word 'conditions' have largely displaced the words 'determine' and 'control' in the study by geographers of environmental effects on man."[37] Some of these tensions are also related to translation: nonscientific words such as "control" or "influence" do not always neatly transpose from one language to another, and Monge M.'s book contains plenty of his own coinings, such as "air temper" and "climatic aggression," which to this day Anglophone readers interpret in slightly different ways (for example, I read the aggression as being part of all climates and influencing evolution in all regions, whereas John West emphasizes the specific "aggression" of low P_{O_2}).[38] But it is clear that *Acclimatisation in the Andes* is making an argument that residents at altitude have a specific physiology, that they are *biologically adapted* to life at altitude in a manner that is hereditary and essentially fixed—in that it takes multiple generations for populations to recover fertility and fitness when moving from mid-altitude to sea level or vice versa (and that this recovery happens only with an "influx of blood" from existing residents).

Despite the fact that there was, at least when *Acclimatisation in the Andes* was published, a vanishingly small amount of physiological and anthropological information about the mid- and high-altitude residents of the Himalaya, Monge M. cites the Sherpa as evidence of his assertion that altitude "makes" a certain kind of person. He claims not only that "the Tibetans" are "agile, strong and hearty," but also that they have an "almost complete insensibility to physical pain."[39] (The idea that Tibetans feel no pain, or at least feel less than Caucasians do—along with the related idea that they do not fear death—has long been used by Western explorers as a semi-justification for the risk of death and injury Sherpa face when assisting White climbers in what is, essentially, a leisure activity.)[40] Despite Monge M.'s claims for homogeneity among mid-altitude populations, as research slowly began to involve Sherpa bodies, the exact opposite was discovered— indeed, the fact that Monge's disease was unreported in Sherpa peoples was part of the evidence physiologists used later in the century to argue that not all dwellers at altitude had the same homeostatic compromises.[41]

After Pugh's preliminary studies on Cho Oyu and Everest, more work was done at Silver Hut (including one serendipitous study of an itinerant monk who turned up on the Mingbo Glacier and astounded the scientists by surviving outdoors without shelter and in minimal clothing).[42] But the first concerted effort to consider Sherpa adaptive physiology did not take place until 1964, during the Himalayan School House Expedition—a venture led by Edmund Hillary with the aim of building several schools and constructing a water supply system for the Nepalese village of Khumjung.[43] On this expedition, Silver Hut veterans Jim Milledge and Sukhamay Lahiri studied the respiratory responses of a handful of Sherpa and compared them with those of sea-level-dwelling visitors. Finding differences not only in the blood (specifically studied to "compare their values with those of the published data of the Andean residents of similar altitudes"), but also in respiratory responses ("The high-altitude residents breathe less than the recently acclimatized lowlanders"),[44] the authors tentatively concluded that Sherpa might have adaptive mechanisms that—unlike those of Andeans—were not available to visitors: "It is possible that complete acclimatization in the sojourners is not the same thing as that obtained in the habitual residents."[45] Moreover, they wrote, "whether these fundamental changes are genetic in origin or acquired over many years or generations of high-altitude residence is one of the unsolved problems of altitude physiology."[46]

Gradually Sherpa were included more and more frequently in physiological studies: blood count analyses were the first (and possibly the easiest form of study to do), but researchers then studied their respiration rates, heart rates, or, like Lahiri and Milledge, the acid-base balance of their blood and cerebrospinal fluid. Nearly all of this work suggested that Himalayan high-altitude residents—particularly Sherpa porters—do not necessarily have the same physiological adaptations to hypoxia that were discovered earlier in Andean populations. Work in this area remained slow (at least compared with the volume of studies on South American bodies), and as late as 1980 it was still a novelty for an American high-altitude researcher to publish the finding that "Sherpas differ from other high-altitude natives: their hypoxic ventilator response is not blunted, and they exhibit relative hyperventilation."[47] But by the last third of the twentieth century, it had become clear that "climatic aggression" could make more than one kind of person at altitude.

There is a counterfactual argument to be made here about how some research programs would have continued had they used Tibetan Man rather than Andean Man as their experimental object—perhaps, if faced with resolutely non-polycythemic blood from Sherpa, their approach to understanding adaptation and acclimatization to altitude would have been different. Well into the late 1960s physiologists were still bemoaning the lack of studies of Sherpa physiology: as British cardiologist Frederic Jackson wrote in 1968, "Unfortunately no [heart] catheter studies have yet been possible in resident Himalayan populations"—which he saw as a huge omission, as it now seemed likely that, in contrast to the more "recent" Andean populations, "thousands of years of residence at this altitude have resulted genetically in a tolerance of hypoxia [to a] degree which the North American and many of the Andean people may not have acquired."[48] The importance of *permanent* residence occurred to other researchers. American (and ex-military) physiologist Robert W. Elsner wrote in 1960 to Pugh to propose some studies on what he termed "mobile" and "stationary" Sherpa, as he had begun to wonder if the South American mining populations, consisting of people who were extremely mobile with regard to the altitude at which they lived, were the same as "more primitive" groups whose "vertical travel" was more limited; perhaps polycythemia was the result of mobility and movement, not long-term residence.[49] (This question can also be read as part of the discourse that represented some people [usually White] as mobile, adaptable, and flexible and others as having bodies that were circumscribed by environmentally deterministic evolutionary processes.) Twenty-first-century explanations of the differences between Andean and Tibetan populations living at altitude also look to mobility and movement as an explanation, and in these genetically informed readings, as Jackson had anticipated, it is the Sherpa, with their "50,000" years of habitation in the Himalaya, who are truly biologically adapted—effectively shaped by "climatic aggression"; the Andeans, with "only" a 9,000- to 12,000-year history at high altitude, are now represented as being more similar to sea-level residents than to the Sherpa, a positioning of Andean Man that would presumably have infuriated Monge M.[50]

Rather than pursuing "what if" histories, it is probably more productive to use these two strands of physiological research to demonstrate the persistent and strongly racialized assumptions about indigenous peoples that threaded through extreme physiology research. Himalayan populations proved much harder to reach with Western medical expeditions due to terrain and politics, but this difficulty was justified or reinforced by an assumption that Nepalese and Tibetan people were less civilized than their Peruvian or Chilean peers. In 1958 Pugh was asked to review a grant application to the MRC made on behalf of the 1959 British Sola Khumbu expedition. Led by a veteran of the 1953 Everest expedition, Emlyn Jones, the expedition aimed to climb Ama Dablam (6,812 m) while also taking ECGs of climbers and Sherpa—for which purpose it wanted £700 from the MRC. Pugh was skeptical: "In my opinion, a cardiological survey of the Sherpa population would be very difficult and time-consuming, because the local people would not be likely to be as co-operative as more educated Western people."[51] In fact, the expedition did get a grant from the MRC, and it did take ECGs of Sherpa porters, publishing the results in 1960.[52] Working with a small sample group of six Sherpa and six "Europeans" (notably, this latter group included a sea-level-born Nepalese climber of "Indian descent"), it found slight differences in the ECGs of the two groups, which the authors thought might be due to acclimatization to hard work and altitude in the Sherpa, or even a "racial difference," but this conclusion was speculative.[53] This is the same study whose published accounts entirely erase the fact that a woman—Nea Morin—was involved in the research.

Women, War, and Warmer Climates

Women, their bodies, and their responses to extreme environments are notably absent from physiological studies until the closing decades of the twentieth century. We have met some rare exceptions briefly in preceding chapters: Mabel FitzGerald, forced to work alone in Colorado as a marginal figure on the Pikes Peak expedition, and Mrs. Sutherland, the mathematical savior of hard-earned acclimatization data from Antarctica. We have seen women deliberately excluded from extreme environments—for example, through the formal and informal bans on women in the Antarctic—and we have seen them excluded as a side effect of the "old boys' network," which required participants in high-altitude expeditions to be vouched for and to have experience (of mountains or of scientific fieldwork). We have seen that even when women did, exceptionally, manage to become involved in expeditionary scientific work, it was still extremely easy for their contributions to be erased through traditional practices of authorship and acknowledgment.

Even when women gained access to expeditions, they were still rarely the *subjects* of physiological studies—an absence that stretched from the field to the laboratory: for example, not one of the three Operation Everest studies of the twentieth century included a female subject. Into the 1980s researchers actively excluded women from experiments because they were "disruptive"—not only, as at field sites, because of fears of sexual or emotional tensions, but because of a perception that women had more dramatic hormonal fluctuations than men, which meant that they could not be used as stable (homeostatic?) experimental organisms. As so many other historians have demonstrated, the assumption that the male body not only was stable and reliable, but was essentially the *normal* form for a human being, is one that has remained stubbornly embedded in modern scientific practices.[54] Just as sea-level-born researchers assumed that sea-level physiology was normal and altitude adaptation a deviation, male researchers figured the female body as atypical.

Of course, by the middle of the twentieth century dozens, possibly hundreds, of women *had* assisted expeditions—including scientific expeditions—to high altitude, and they were not (just) laboratory assistants, supportive wives, instrument makers, or "human computers," but were actively engaged on the expeditions themselves. These women, however, are doubly invisible—not only female, but non-White. Pugh wrote about an encounter with one such woman on the trek to Everest Base Camp in 1953:

> April 11th: Area of bad terrain. One sherpani slipped down a slope with her load round her neck. I happened to be by & went to help her. All the time relaxing her and fixing her load, she kept her head turned away, so I did not see who it was. I supposed she was ashamed at having fallen.[55]

Despite, or perhaps because of, the personal, almost tender, nature of this interaction, the presence of Sherpani on the expeditions is a note in Pugh's diary—it is not a fact one would gather easily by reading his official publications.

The gender bar at high altitude lasted longer for indigenous high-altitude peoples than it did for almost any other ethnicity.[56] As an elite vocation, high-altitude portering was an exclusively male activity until the very end of the twentieth century; the first Nepalese woman to reach the summit of Everest did not do so until 1993—and the first to reach the summit and survive the descent not until 2000.[57] But on the long treks to Base Camp—so essential for the slow acclimatization of White climbers—women carried food, shelter, fuel, and scientific equipment; they also set up camps, cooked, fetched water, and assisted with experiments and scientific observations. Despite these roles, they remain anonymous "Sherpani." Likewise, it is as anonymous non-White participants that we come across women's bodies most commonly as the subjects of, as well as the assistants to, extreme physiology research: the influence of anthropologists on the study of extremes of heat and cold meant that women's bodies were included in studies of *racial* adaptations to non-temperate environments, even as they were excluded from physiologists' acclimatization studies in laboratories and on expeditions.

There is a gap in our understanding of the relationship between acclimatization, adaptation, racial science, and evolution. The positive-to-negative narrative of inevitable White survival versus White racial degeneration that runs from the late eighteenth to the early twentieth century has been clearly articulated, but it is less clear what happened to this relationship through the rest of the twentieth century. It is not the purpose of this chapter—or of this book—to entirely fill in this lacuna in our understanding; but what it can contribute is a case study from extreme physiology that gives an indication of the relationship between adaptation science and evolutionary theory. As part of the larger argument outlined here, it is clear that physiological work in the twentieth century has been overshadowed—in historical as much as in biological study—by the arrival and eventual dominance of molecular biology and genetic studies. From early surveys of blood groups, which shattered existing understandings of race and race fixity, to early twenty-first-century studies of mtDNA as a possible source of altitude adaptability, these sciences offered new explanations and models for human pasts and human futures.[58] While some scholars, particularly Warwick Anderson, have made a strong case that physical anthropology—and therefore physiology—were extremely important in shifting understandings of race and ethnicity, physiology does seem to have become a minority player in this field of study.[59] One exception, however, is in the study of acclimatization and adaptation to extreme environments, where physiology remained a core specialty. There is considerable overlap between the people working on the bodies of explorers and those working on the bodies of indigenous residents of tropical, high-altitude, and Arctic environments; likewise, there is a constant dialogue between theories about temporary White acclimatization and long-term non-White adaptation. Consequently, a closer look at acclimatization science here is the start of an alternative story about biomedical research and race in the modern period, one that centralizes holism and fieldwork.

While we have already seen a clear and repeated crossover between studies of high altitude and extreme cold, it is in the study of indigenous peoples that the links between *hot* climates and other extreme environments become most obvious. Blood, the core of this chapter, remains a key substance; the blood-regulating homeostatic responses of the body to high temperatures and UV radiation were often seen as complicating and confounding factors in acclimatization studies. For example, through the 1920s a handful of European researchers—most prominently the German physiologist O. Kestner—speculated that solar radiation could cause polycythemia; the "tropical" sun was believed to increase red blood cell production in the "normal" person (normal here to be read, of course, to mean a White European).[60] In 1926 Dr. D. V. Latham of the East African Medical Service climbed Kilimanjaro, making observations of his and his companions' blood; he published results showing they had higher hematocrits than "the Everest climbers for corresponding altitude," indicating that Kestner was right and high UV radiation provoked red blood cell production.[61] Through the 1920s, '30s, and '40s, studies in the laboratory and the field seemed to suggest "that the human body contained a greater volume of blood in a hot environment than in a cool one."[62] By 1949 studies on British naval personnel at Cambridge University led one physiologist to speculate that the movement of blood between the surface and the inner parts of the body, as well as changes in the volume of circulating blood, might play a part in the processes of acclimatization to both cold and heat.[63]

As this 1949 study of sailors hints, during the first half of the twentieth century interest in acclimatization to hot climates was often shaped by military priorities, or at least by military funding. Work motivated by the Second World War established the role of increased solar radiation in dehydration, sunstroke, fatigue, and cancer, if not in increased blood cell production. In 1947 the American physiologist Edward F. Adolph edited *Physiology of Man in the Desert*, probably the first heat-adaptation textbook published in the twentieth century; in what is a review of largely US military-related research, the volume's authors are emphatic that no serious work was conducted on heat survival (including such basic issues as water requirements) until 1941, when the US joined the Second World War and faced the prospect of desert combat.[64] While this claim ignores some of the nineteenth-century colonial studies, it is true that military interest in hot climate survival drove a significant increase in such research after 1940; it is notable how often the US military, in particular, found funds for Anglophone research into these topics. It was the US Air Force that partly funded Allan Rogers's analysis of his TAE study (and therefore paid the salary of Mrs. Sutherland), and it was the US military that funded, quite generously, studies on heat adaptation among various indigenous peoples in Australia and other Pacific nations. The "significance" of one such study in 1961, as claimed in the research grant proposal, is specifically that the bodies of indigenous people can help in understanding the problems of European settlers:

> The long term adaptation of desert nomads to the hot and dry environment should yield additional information on the habituation possible in Europeans and the part of high water intake in their adaptation to desert living.[65]

But positive statements in grant proposals do not always translate into positive findings. An international team put together a project in the early 1970s, under the aegis of the Human Adaptability theme of the IBP, to study the heat regulation responses of people in New Guinea. The study, which used heated beds and tested both men and women, produced a very ambiguous answer to what appeared to be a simple question: Do long-term residents of hot climates have more efficient sweating mechanisms than residents of cooler climes? Many previous studies (and most personal experience) had shown that the immediate and obvious response of an unacclimatized body to a raised environmental temperature is sweating, but some work had suggested that ethnic groups indigenous to high-temperature regions had lower rates of sweating than White visitors. This was an Andean versus Sherpa question: Are indigenous people able to survive in extreme environments by doing more of the same thing that White bodies do (which is a good sign for the potential of long-term White acclimatization), or do they have an entirely different survival mechanism (which, if hereditary, is inaccessible to the White soldier or settler)? Unfortunately for the funders interested in the acclimatization of temperate peoples, the answer appeared to be that whatever the mechanism was in permanent residents, it was not simply an up-regulation or a tweak to the mechanisms already observed in White bodies and could not, therefore, be translated into long-term acclimatization schemes for Europeans (in other words, it was a Sherpa-style adaptation, not an Andean-style acclimatization).[66]

This 1970s study was a joint enterprise between the UK's National Institute for Medical Research and the School of Public Health and Tropical Medicine in Sydney—highlighting the fact that internationalism is a key feature of extreme physiology studies, as has been repeatedly shown in the preceding chapters. Australia and, to a lesser degree, New Zealand were important contributors to studies of extreme physiology: as well as providing high-altitude climbers for British expeditions, these countries were important ports used on the way to the Antarctic, and they also provided academic and institutional bases for key researchers working on the physiology and anthropology of populations living—temporarily and permanently—at high temperatures. The joint New Guinea study is one example, but Australian researchers in particular also turned to the bodies of their nation's own indigenous peoples to try to answer questions about acclimatization, survival, and evolution.

It was a widely held opinion in the early twentieth century that the Australian Aboriginal peoples would become extinct within a few generations—a prospect that some White Australians regarded as a positive outcome, and which was deliberately hastened by policies such as the removal of mixed-race children from Aboriginal families. At the same time, many looked anxiously at the attempts to settle Queensland, and debate about the possibility of creating a healthy, non-degenerate "tropical White" race in that area (and others) was still live through the start of the Second World War. The classic text on the topic, and a very uncomfortable read, is the historian and geographer A. Grenfell Price's *White Settlers in the Tropics*.[67] Grenfell Price rejects environmental determinism, but does so in order to promote the possibility of White settlement across the globe, and he cheerily considers the extinction of "lesser" races, citing the degradation of the Negro as the biggest threat to White success in hot climates.[68]

There were dissenting, or at least more sympathetic and egalitarian, voices. One good example is Walter Victor Macfarlane, a New Zealand–born physiologist who spent much of his working life based in Adelaide, Australia. An animal, rather than human, physiologist by training, he worked extensively on (and with) various Australian Aboriginal groups as well as on the heat survival abilities of African Bantu groups and New Guinea Highlanders, among others. He was also, as his correspondence shows, a researcher who acted as an international hub, exchanging data, ideas, drafts, and support for grant proposals with scientists across the globe. (The contingency of these relationships is well illustrated in a letter he wrote in 1980, a year or so before he died, to Dr. Philip G. Law, an Australian Antarctic explorer and leader of many Australasian Antarctic expeditions: "It was a pleasant chance that put us side by side on the plane recently.")[69] What Macfarlane's work suggests is that first, not all populations indigenous to hot countries adapt to heat in the same way—as was also being established for high-altitude populations—and second, in the specific case of Australian Aboriginal populations, these were *tropically adapted* peoples surviving in the dry heat of the Australian desert using largely *cultural* adaptations. That this was the case seemed evidenced by the effect of encroaching Westernized lifestyles—while many people suggested that Western diets could lead to the eventual extinction of indigenous peoples (through rising rates of cardiovascular disease and tobacco- or alcohol-related illnesses), Macfarlane pointed out that even water drinking (and sweating) habits could be changed within an individual lifetime: comparing two indigenous groups with differing levels of contact with White culture, he said that "when it came to electrolyte handling, however, the Pintubi were much nearer to the White man than the Nadajaras. Three years of flour, sugar, salt and canteen foods had apparently had their effect."[70]

Repeatedly throughout his career, Macfarlane expressed his opinion that acclimatization to heat was not likely to be a hereditary feature:

[While] there are group differences in this sort of thing . . . whether they rise basically from race or from habituation to hot or cold environments is something which will [require] another sort of investigation . . . *I doubt whether race is the answer* [my emphasis].[71]

This tension between short-term and long-term, between cultural and technological, and between biological and evolutionary adaptation has been seen in the two preceding chapters, although in a very different context: the repeated failure to find significant biological adaptations to cold gradually diminished the importance of physiological research into Arctic populations (see chapter 3) while simultaneously increasing the importance of "bioprospecting"—that is, the borrowing of cultural and technological practices from populations in the Far North to aid survival in the Arctic and Antarctic (see chapter 4).

While Macfarlane's work is evidence that the dichotomous assumptions that linked hot versus cold, civilized versus uncivilized, and technology versus biology were in no sense unanimous across the field of physiology in the middle third of the twentieth century, his views still represented only a minority opinion about racial adaptation and White acclimatization. The idea that human races come in "primitive" and "civilized" types is frequently encountered in the writing of extreme physiologists. As an example, in a write-up of the International Biological Expedition to Antarctica (IBEA)—which was unusual in the late 1970s for concentrating on cold acclimatization with its attempts at "pre-adaptation" using cold baths and such—the expedition leader, Jean Rivolier, is explicit about the differences between "primitive" and "civilized" people: "Man's ability to adapt has enabled him to live and work successfully in hostile climatic environments throughout the world"; and while many of these successful adaptations have been technological or behavioral adaptions, they "may also include alterations in physiological responses."[72] There is no mention of the "Eskimo" or other circumpolar or cold-adapted people in this category of those who "may" have "alterations in physiological responses"; instead, he lists Kalahari Bushmen, Australian Aboriginals, Alacaluf Indians from the Andes (Tierra del Fuego), and the diving women of Korea and Japan. It cannot be a coincidence that all these groups were widely described as nomadic, "untouched by civilization," or "primitive," or, of course, were *women*. "On the other hand," Rivolier writes, we can compare this with what he calls "classical adaptation, as seen in temperate zone dwellers visiting polar climates, [which] is epitomised by successful *technical and behavioural stratagems* which virtually, but not entirely, eliminate climatic variables as a direct stress [my emphasis]."[73]

Time and again, the Eskimo, and other cold-region dwellers, were specifically excluded from discussions of biological adaptation, and while in many cases their exclusion reinforced a hierarchy that put tropical peoples below those of temperate regions, by midcentury it was also the source of criticisms of environmentally deterministic theories. It is not a coincidence that many of the scientists who challenged determinist thinking in racial science and evolutionary analysis were researchers who had spent time in the

Arctic or Antarctic. As examples, we can take Laurence Irving and Per Scholander, who in the 1950s and '60s critiqued two "rules" about anatomical proportions and environmental stress.[74] The first of these, Bergmann's rule, suggests that individual animals of the same species tend to be smaller in warmer climates and larger in colder ones; the second, Allen's rule, similarly argues that the extremities—notably limbs—show the opposite tendency, becoming smaller in colder climates. Both Bergmann's rule and Allen's rule are based on mathematical calculations of heat loss and the surface-area-to-volume ratio of living organisms, and they were applied by physical anthropologists to humans not only as a mechanism to explain differences, but also as a proposed way of identifying human subspecies, or races. While such measurements may seem trivial, we should recall the widely quoted suggestion from the arch-eugenicist and racist Charles Davenport that one scientific justification for racial segregation was that "mulattoes" (mixed-race individuals) have unfortunate and unhealthy body proportions because of limb-length incompatibilities between races.[75]

Irving and Scholander used their work on animals—in this case, field-work and studies of living organisms, not mathematical extrapolations—to dispute the two rules and, in particular, their application to human races. Like many other authors, they stated explicitly in their writings that Eskimo survival was *technological*, not *biological*; as Scholander put it in 1955, "In the Eskimo the main adaptation lies not in physiology, but in age-long experience and technical skill."[76] So far, like many other authors on human adaptation; but what was unique about these two authors was their explicit recognition of similar behavioral adaptations in animals—if even dogs could exhibit cold-weather behaviors that invalidated the determinist, rigid anatomical rules of Bergmann and Allen, it seemed ridiculous to suggest that the ingenious human being would have a body so irrevocably shaped by a temperature cline. From Arctic and diving animals, Scholander went on to study the Alacaluf Indians and sleeping Australian Aboriginals—both ethnicities cited by Rivolier, above, as having biological adaptations—sometimes drawing conclusions strikingly different from those of the determinist evolutionists who not only studied the same peoples, but even went on the same expeditions.[77] But while physiologists like Macfarlane, Irving, and Scholander offered counterexamples to environmental determinism, the general consensus within Western biomedicine, at least into the last third of the twentieth century, was that technological and behavioral adaptations were the features of peoples of cold climates, while biology and instinct dominated the lives of those of hot regions. As the IBP's Human Adaptation theme leaders said, "Because of the absence of marked physiological differences between Eskimos and whites, it was concluded that the principal adaptation of these Eskimos to their climate is technological."[78]

The conflict over environmental determinism was a live and ferocious debate, played out across animal and human physiology and evolutionary theory, in private letters and in articles in *Nature*. Yet while it drew on, and in some circumstances influenced, research in extreme physiology and expeditionary science, it seems to have had less impact on researchers looking at the body responses of explorers and visitors in the Arctic, Antarctic, and high-altitude regions of the earth (and there is much less evidence of animosity and debate within this area). I would suggest there are three main reasons for this: The first is the effect of the "frozen field" itself, which created a relatively tight-knit core of researchers in human physiology and exploration who, while they read and borrowed from broader work on evolutionary theory and animal physiology, did not get involved in the serious fractures and debates of those fields. Second, most of these researchers were medical doctors or physiologists with a commitment to "whole-body" work, such as Pugh and Pace; while they were interested in the fields of molecular biology and genetics, their focus was on individual bodies, on organs, systems, and experiments, not on gene pools or mathematical models of populations. Third, and finally, with the possible exception of studies on some South American populations, this research was dominated by studies of the bodies of Western explorers and scientists. The twentieth century was half over before Sherpa porters were recruited into even the most basic studies of respiratory or cardiovascular function, while *physiological* studies of circumpolar and tropical or desert-dwelling peoples were patchy at best. Indeed, this lack of research was acknowledged explicitly at the formation of the IBP, which, with its Human Adaptability theme, aimed to survey and study indigenous peoples as a matter of urgency before "primitive" communities became too influenced by "civilization." (While some of this work was conducted in temperate zones, the IBP specifically singled out populations living in "extreme" environments as priority areas for study.)

Field Sweat versus Laboratory Sweat

While the decade-long IBP scheme began to systematically investigate hereditary adaptation and environmental determinism, extreme physiologists continued their work on the acclimatization of non-indigenous peoples to hot, cold, or high-altitude environments. One ongoing, but not particularly successful, theme in this research consisted of attempts to "pre-acclimatize" individuals to these environments. As we saw above, the possibility of pre-acclimatization of sea-level residents to altitude by blood packing or hemodilution was certainly discussed, although it turned out to be impractical, and of ambiguous value, on the mountainside. As late as 1977 the IBEA was trying to pre-acclimatize explorers to Antarctica by using cold baths and other stressful treatments. Such attempts to pre-acclimatize subjects from temperate climates in order to give them a "head start" in adaptation—or, in other circumstances, to deliver troops at maximum physiological efficiency to fields of conflict—suffered from one of the problems that had plagued altitude researchers in the previous century: uncertainty about whether artificial simulations in laboratories or climate chambers could really mimic the natural embodied experience of a particular (changeable) climate. (This problem also had clinical corollaries, as there was a parallel debate about whether frostbite or frostnip "simulated" in the laboratory was really the same as that experienced in the field.)[79]

We have seen in previous chapters not only how barometric, climate, and wind chambers were used in extreme physiology research, but also how their results were always tempered by, and used in comparison with, real-world experiences and experiments. While these chambers were often acceptable proxies, the "real" test of a piece of equipment, a theory, or an explorer's body was Antarctica, the Arctic, or a mountainside. The same process was happening in heat adaptation research: through the early twentieth century some evidence from heat chambers suggested that repeated, prolonged, or extreme exposure to heat altered physiological responses, perhaps as a form of acclimatization, usually by making the body's cooling reflexes more efficient. But the question remained whether artificial acclimatization in heat chambers necessarily mirrored the experience of living in tropical or dry-heat areas. The military interest in this area of research is demonstrated by the fact that the first systematic attempt to directly compare artificial with natural acclimatization was a 1950s study jointly funded by the UK's Medical Research Council and the British Royal Navy, which compared recruits in heat chambers with staff in Singapore.

The MRC-Royal Navy studies were co-led by J. S. Weiner, a South African–born British researcher who would go on to become the convener of the Human Adaptability theme of the IBP. The project's review of the existing literature turned up many previous heat chamber studies, all showing similar acclimative responses to heat in Europeans and North Americans (essentially, increases in the sweating response and in the amount of sweat produced, alongside reductions in body temperature and increases in heart rate, upon exposure to higher temperatures), although as Weiner and his co-authors pointed out:

It has been assumed that [these adaptations] constitute . . . the adaptation to high environmental temperatures that occurs naturally. . . . This assumption is implicit in most of the laboratory work done . . . but, as far as is known, it has never been rigidly demonstrated.[80]

This study's conclusion was that artificial acclimatization was "identical" to natural acclimatization—suggesting that pre-acclimatization to heat could be a useful technique for the armed forces and for explorers. Further, it seemed to confirm that acclimatization to heat was at least in part a physiological response (rather than a technological or behavioral one). The question left unanswered was whether there was a difference between this short-term acclimatization and the long-term, hereditary adaptation of indigenous peoples; indeed, as we saw above with the New Guinea study, this question remained unanswered into the 1970s.[81] On finding that indigenous peoples present a different response to heat, the New Guinea researchers suggested that there is a fundamental difference between the biology of indigenous peoples and of Caucasians. They suggested that the sweating response is a temporary or short-term one that in the "longer term is replaced by other adaptations reducing the need for sweating," but remained ambivalent about whether the "longer term" adaptations are racially fixed (environmentally determined) characteristics or something to which White bodies could aspire. (It is notable, too, that Macfarlane's work is cited in this study as a rare counterexample to the then-dominant theory that adaptation to heat is racially linked and evolutionary in origin.)

Research into the acclimatization of White bodies to hot climates was also relevant to the explorers and scientists who traveled to high-altitude and circumpolar regions in the twentieth century. Indeed, observations about the relative demands of man-hauling and dog-sledding (discussed in previous chapters) finally led one physiologist to study man-hauling Antarctic explorers—using a swallowable temperature recorder—for evidence of acclimatization to *heat*.[82] For most cold-weather researchers, what made heat acclimatization studies interesting was their focus on the behavior of a body fluid—in this case not blood, but sweat. As mentioned above in relation to hemodilution, dehydration was a serious factor in early disasters at the poles and in failed high-altitude expeditions. Dehydration was a problem for explorers at high altitude not just because of the sweat of exertion, but also due to increased water loss caused by increased respiration rates (which meant that artificial oxygen supplies, if well designed, could actually prevent thirst as well as fatigue). Hydration was a challenge for Arctic and Antarctic explorers, too, because despite the wide availability of ice and snow, using them for hydration demanded a thermal compromise: either one expended effort transporting fuel to melt them, or one used up one's own body heat to do so. Sweat was also a complicating factor for clothing design—as we saw with Allan Rogers's study and will see with more of Pugh's work in the next chapter, the thermal properties of clothing change dramatically if it is dampened with sweat, or if wearers untuck or unzip clothes in response to the body heat they produce by doing hard work.

The intertwining of these research strands—the fact that heat chamber studies in Singapore could influence sock design in Antarctica—mirrors the international webs of interest, connection, and intellectual heritage outlined earlier in this book. Few scientists or explorers remained within one environment; to take Pugh as an example, while he never engaged in an expedition to a tropical or desert environment, he did undertake research into hypo- as well as hyperthermia and into the dehydration caused by high temperatures and fatigue. These broad interests are self-fulfilling prophecies for expeditionary scientists, who through the twentieth century make a clear case for what I term *whole-body physiology*. Just as the early arguments over altitude acclimatization focused on the ability of models such as the barometric chamber to reveal truths about practice in the field, expeditionary physiology relied on multifactorial studies to try to understand the complex, and variable, homeostatic responses of the human body to changes in its environment. Much earlier than ecosystem thinking began to dominate in the environmental and ecological sciences, a form of "internal ecosystem" research persisted in the biomedical sciences, working alongside (and occasionally challenging) more focused, if not reductive, genetic understandings of human physiology.

Evidence of this holistic perspective can be seen in repeating patterns within physiological research: for example, in a mirror of the late nineteenth-century debates about whether mountain sickness was caused by altitude alone or by altitude and fatigue, mid-twentieth-century researchers debated whether heat acclimatization could be achieved by exposure to heat alone, or if physical exertion was a necessary stimulus to the essential homeostatic changes. Early studies on the effects of heat on the human body had been clearly confounded by these interactions with fatigue. Through the 1930s several Harvard Fatigue Laboratory projects were conducted on the effect of heat on work and productivity, and while these were studies on labor efficiency rather than on physiological adaptation, the possibility of acclimatization was a complicating factor. One such study, on Black and White subjects in Mississippi, found that Black subjects "were able to perform a standard walk in the heat with the smallest rectal temperature rise"; the

researchers suspected that this might be due to the fact that the Black subjects were far more accustomed—read, *acclimatized*—to long, hard work in
the sun than the White sharecroppers and laboratory workers who also took
part in the study.[83] It is notable that these researchers do not assume that
African Americans have a genetic or racial advantage or a hereditary adaptation to working in the heat, but rather consider the possibility of acclimatization within an individual's lifetime, during which physical fatigue and
work rate are crucial co-components of acclimatization.

The apparent complexity of these extreme physiology studies—as joint
field and laboratory projects in complex environments, considering multifactorial influences, often with interdisciplinary and international teams—belies how narrow they often were. As we might expect from twentieth-century
biomedicine, these projects took the White male body as the unquestioned
default for their studies. Indigenous bodies, although potentially informative,
were *deviations*, showing adaptations or changes from the "norm." Of course,
some individuals pushed back against those definitions—in particular, South
American researchers made a case for high-altitude populations to be considered the normal and sea-level populations the deviation; this argument is in
part a philosophical exercise, but has a root in fetal biology:

> During intrauterine life the arterial oxygen tension in the fetus is 20, which
> would correspond to an atmospheric oxygen tension of approximately 61—
> that is, to an altitude of about eight thousand meters above sea level. There-
> fore, at birth, wherever it occurs, the fetus passes from a hypoxic to a better-
> oxygenated environment.[84]

(It is notable that after altitude, Barcroft's greatest research output was on
fetal life—he even used the term *Everest in Utero* to describe the hypoxic environment of the womb.)[85] In particular, the Peruvian physiologist Alberto
Hurtado maintained that the Andean Man of Monge M.'s construction was
a valid norm for study because there were "qualitative differences between
a man born and reared in the hypoxic environment and one who has been
subjected to it temporarily," and he argued that a distinction should be
made, in research publications at least, between what he termed "natural"
and "acquired" acclimatization.[86]

Women, too, made a case for their biology to be regarded as normal. Despite all the challenges, women were present at high altitude and in spaces
of extreme cold and heat by the 1950s; but while they were taking part in
expeditions and participating in scientific work, what they were not asked
to contribute was blood, or sweat, or data. None were invited to pedal the

ergometer at Silver Hut or to be strapped into an IMP in the Antarctic. Their absence did not go unnoticed: at the Symposium on Polar Medicine held in Cambridge in late 1959, the all-male attendees did feel "that the time had come to observe the reaction of women as well as of men," although none seem to have taken any action to further this goal.[87] The absence of experimental evidence relating to female performance did not stop speculation, however. Nea Morin mentions in her autobiography that she had heard a theory that women might do better in anoxic environments because their smaller brains required less oxygen.[88] While she seems somewhat dismissive of this idea, she takes more seriously the conjecture that women, having higher levels of body fat, might be systematically more resistant to the "climatic aggression" of low temperatures.

While physiologists were rather slow to study this possibility for White women, the bodies of non-White and indigenous women attracted more scientific scrutiny. In particular, the advanced free diving skills (that is, diving without supplemental oxygen) of the Ama of Japan and the Haenyeo of Korea[89] attracted the attention of several physiologists in the middle of the twentieth century, as they raised the possibility that women might have physical advantages in cold, low-oxygen environments. These diving women could spend several hours in cold water, stay submerged for several minutes, and descend to 24 m (80 feet) or more in pursuit of octopus, bivalves, sea cucumbers, and other marine animals. They were famed for continuing to work through pregnancy and for having careers that extended into their eighties. Their practices were, to the observing scientists, a complicated mix of technique, adaptation, and tradition, which even into the 1970s proved difficult to analyze: their techniques of hyperventilation had obvious laboratory-proven value,[90] but why did they limit their final breaths before diving? Was it to reduce buoyancy? And was there any truth in the suggestion that their pre-dive whistle "protected the lungs" somehow?[91] What did appear certain was that the diving women were exposed to more cold stress than any other population on earth (the authors of one study made the point that "Eskimo" populations make use of technology—clothing—to prevent such extreme exposure),[92] and that one reason they were diving *women* was because of the enormous insulating importance of subcutaneous fat, as "[women] are more generously endowed with this protection than men are."[93]

Indeed, taken objectively, women's bodies seem far better suited to cold-weather exploration than men's—not only because of their body fat, but also because facial hair is an enormous inconvenience (even a risk) both at high altitude and at the poles. Stefansson insisted in his *Arctic Manual* that an explorer "should always be clean shaven" because "if you wear a beard, the moisture of your breath congeals on it," creating an uncomfortable "face mask": "If you try to thaw such an ice mask with your hands, you soon find that you have to choose between a frozen face and frozen fingers."[94] At altitude this issue was even more serious, as even rough stubble could interfere with the fit of oxygen masks, which at low temperatures iced up rapidly, preventing movements of the neck and head; "more serious[ly] . . . the clothes freeze to the beard, and lose their warming property."[95] These problems interfered with the practices of experimenters, making male bodies very "disruptive" to scientific research: "The attempts to determine oxygen consumption during the test cold exposures at Mawson had failed because of uncontrollable leaks caused by the subjects' beards."[96]

Conclusion: Blood Sports

While White women's bodies were relatively absent from extreme physiology studies in both field and climate chamber, this was not the case for non-White female bodies, which were widely scrutinized as part of anthropological, genetic, and evolutionary studies. As this chapter has outlined, the relationship between such studies of long-term hereditary and evolutionary adaptation (in both men and women) intersected in complicated ways with the much more short-term focus of expeditionary scientists. While behavioral, technological, and physiological responses to extreme climates were of interest because they offered options for White acclimatization, evolutionary and hereditary responses were more complicated to understand, and perhaps of less immediate use to explorers (or their funders in the European and North American military). In the nineteenth century those who claimed that non-White bodies were better adapted to hot, cold, or high-altitude environments had to articulate that hereditary advantage within a social and scientific context that maintained the White body as superior. In the twentieth century those conversations remained difficult to frame—while strong racist presumptions remained within genetic, physiological, and anthropological work well into the twentieth century, researchers also had to juggle the dramatically shifting popularity of evolutionary determinism (illustrated here by the ambivalent introduction to Monge M.'s *Acclimatisation in the Andes*) and the changing definitions of (even outright skepticism of the existence of) race and racial groups.

This chapter has only been able to gesture toward the nature of the relationship between short-term acclimatization and long-term adaptation science. What it has shown is, first, that assumptions about the role of environments and the capacities of indigenous peoples—from the relatively late recognition that "climatic aggression" could produce more than one kind of acclimatized body to the extraordinary tenacity of the assumption that hot-climate adaptation was biological and cold-climate adaptation was intellectual—have shaped physiological research. Second, it has emphasized the "whole-body" or multifactorial nature of studies that had to consider heat production in Antarctica, dehydration on Everest, and cold overnight temperatures in the Australian desert. In particular, fatigue appeared time and again as a complicating factor: fatigued participants (and this included the mentally fatigued and distressed, such as those who had undergone a series of unpleasantly cold baths) responded more strongly to climatic stresses and performed worse at high altitude than the unfatigued; at the same time, multiple studies suggested that active subjects acclimatized more quickly to heat and altitude than did sedentary ones (while, of course, hard work made cold acclimatization *less* likely, by preventing bodies from experiencing the full shock of the Antarctic or Arctic cold). Chapter 6, which concludes this book, will look more closely at the complicated role of fatigue in extreme physiology. Before that, there is a story that needs to be finished here, and it is the story of blood.

L. G. C. E. Pugh has been featured here mostly as a physiologist of extreme spaces, but he also maintained a strong interest in research into more mainstream sports medicine. This research, unusually, included women's bodies: through the late 1960s, while he was still analyzing data from the Himalaya, Pugh was also working with the British Women's Olympic Skiing Team to provide fitness tests, training analysis, and advice on competing at altitude.[97] The non-mountaineering sporting world had become particularly interested in his skills in the 1960s, after the announcement that the 1968 Olympic Games were to be held in Mexico City, at mid-altitude. Pugh was recruited by the British Olympic Association to travel with a team to Mexico in October 1965 to conduct studies of the effects of altitude on the performance of elite (male) athletes.[98] The increasing economic and political significance of international sports provided a means for extreme physiology research to find new funding, new human subjects, and new areas of interest, which included finding ways to improve performance at sea level, not just at altitude. By the 1970s polycythemia was only disputably an adaptation—possibly even a *mal*adaptation—to altitude, but it was clearly an advantage to physical performance at sea level. In 1963, as Tom Hornbein was speculating about the possibility of "hastening polycythaemia" and outlining the reality of the failure of drugs and the dangers of transfusion, he suggested that the best hope was that "someday if erythropoietin is ever synthesized, the body may be able to do the job itself."[99] In fact, the first blood-doping scandal was caused by a more primitive method of hastening polycythemia: in 1984, at the Olympic Games in Los Angeles, the USA Cycling Team broke a seventy-two-year medal drought by taking home nine medals—and a few months later the story broke that some team members had benefited from autotransfusions during the games.

The possibility of blood packing by autotransfusion—removing an individual's blood, storing it, and then transfusing it back into them closer to the event date—had been widely (and openly) discussed in medical journals and sports training circles, and even in sports magazines such as *Track and Field News*.[100] Studies and opinions were mixed, so, if nothing else, the 1984 experience of the USA Cycling Team seemed to act as confirmation that transfusion worked. It was also not, strictly speaking, cheating. No provision against transfusions existed in the International Olympic Committee's rules. While various restrictions on doping had been in force at Olympic events since the marathon in London in 1908, the IOC had not created a list of "banned substances" until the 1960s, and neither blood nor EPO was on the list as of 1984.[101] One reason for this was that the IOC was unwilling to ban substances for which there were no tests (which seemed like an exercise in futility); another reason was that there was a certain ambiguity about blood doping—blood was, after all, an athlete's own body fluid; it could be said that the technique merely mimicked the effect of altitude training, which, although completely legal, could be expensive or inconvenient for some athletes, who might then find themselves at a competitive disadvantage against their Alpine- or Kenyan-trained opponents.

As we saw with oxygen in chapter 2, in the ethics of sport and exploration, the boundary between good preparation, sensible use of technology, and "cheating" was a porous and complicated one, formed by national priorities, class and gender identities, and climbing traditions. The same was the case for doping regulation in international sports, a contentious, expensive, and frequently arbitrary practice that has provoked a huge amount of academic scholarship.[102] Like assistive climbing technology, doping policy has huge ethical consequences: by the late 1990s, multiple athlete deaths had been blamed on EPO use, particularly in cycling. When dealing with people intending to perform to the limits of their physical capacity, international sporting organizations and expeditionary biomedicine had to seriously consider the risk that the participants would die. Unlike sporting events, expeditions posed a risk of death not just to the direct participants—the climbers, explorers, and scientists—but also to a broader community of supporters, technicians, and, most obviously, guides. Seven Sherpa died as a result of George Mallory's attempt on the summit in 1922, and while he clearly felt a responsibility for their deaths—the first recorded on Everest—that did not stop him from making another Sherpa-supported attempt in 1924. In the following, and final, chapter, we will consider these two themes: the ways in which extreme physiology bled out into other areas of research, from incubators for newborns to rations for NASA; and how participants dealt with the ethical and emotional challenges of their occasionally fatal work.

SIX

Conclusion: Death and Other Frontiers

The human subjects of extreme physiology investigations often have to do extraordinary things. In the mid-1960s one such guinea pig literally walked in a dead man's shoes; asked by the British physiologist L. G. C. E. Pugh to wear clothes taken from a dying man, he endured a cold shower and the chills and wind produced by a climate chamber. Shivering in his outfit of jeans, wool jersey and shirt, and hooded anorak, the twenty-one-year-old volunteer also worked on a bicycle ergometer, and thus helped Pugh produce data that later became a note in the prestigious journal *Nature*.[1] Pugh's key finding was that a combination of wet *and* windy conditions dramatically reduced the thermal properties of clothing, meaning that outfits thought adequate for certain temperatures offered nowhere near enough protection for the inclement weather of the British Isles. Jeans proved to be a particularly poor choice for walkers in bad weather, and overall the insulation value of the outfits studied, when wet and exposed to winds of just 9 miles per hour, fell by 85% or more.[2]

Pugh had obtained the clothes by writing to the families of three young men who had died taking part in the 1964 Four Inns Walk. This walk had been organized yearly since 1957 by the local Rover Scout group, and in 1964 it attracted more than 200 teams ready to take on the roughly 40-mile (65 km) walk through a northern section of the Peak District in the UK. Partway through the event the weather turned, and fewer than 30 teams successfully completed the route; several walkers had to be evacuated or rescued, and only 22 of the original entrants actually completed the full route. There were three fatalities—G. Withers, J. Butterfield, and M. Welby—all of whom died of the effects of exposure, one in hospital, the other two in the wild; the rescue operation took three days to locate both bodies. They were by no means the first such fatalities. Around the middle third of the twentieth century, an unprecedented number of people were seeking access to the wilder spaces of the British countryside. Rising wealth, increasing leisure time, and better transport combined with sociopolitical movements—such as the Kinder Scout mass trespass (in support of access to open spaces) and the offshoots of the interwar physical culture movement (e.g., Outward Bound)—to bring young people to the mountains of Wales and Scotland and to the Lake and Peak Districts. While serious incidents were far from commonplace, they made headlines when the victims were children: in the early 1960s, prior to the Four Inns disaster, an eighteen-year-old man had died on an Outward Bound expedition from Eskdale in the Lake District, and a sixteen-year-old boy had died while training for his gold Duke of Edinburgh's Award in the Radnorshire Hills in Wales.[3]

Death is never far from the practice of extreme physiology: whether one is retrieving a lost mountaineer's ice axe on a scientific expedition, remembering friends at a snow-covered cairn in Antarctica, or squeezing into the vest taken from a dead teenager, there is no escaping the dead and the dying when one studies the limits of human survival. While fatalities have been threaded through this book, this final chapter will take a harder look at the human costs of, and the ethical debates over, extreme physiology. At the same time, it will look at the positive outcomes of this research—after all, the results of Pugh's work on hypothermia ended up completely changing the advice given to people in some emergency situations and underlies the life-saving advice given to even casual walkers in the UK today. This chapter will also consider the overarching claims of this book: that extreme and exploration physiology is a unique—and uniquely deadly—form of science, and yet still offers us important insights into the practices and values of scientific work in the twentieth century and, further, that these insights are an important corrective to the existing "big-picture" stories.

Exhausting All Options

The deaths on the Four Inns Walk became case study number 23 in a survey of serious incidents on British mountains that Pugh published in 1966.[4] Pugh used these extremely varied cases—ranging from a middle-aged Indian tourist who got lost in Snowdonia to a party of Royal Marines commandos who were taken by surprise by a weather change in Llanberis, Wales—to make a series of recommendations not only about clothing (no jeans, better attention to waterproofing), but about *behavior*. One of the problems of the Four Inns disaster had been that, once they were cold, pre-hypothermic, and low on energy, many participants continued to walk in the hope of finding shelter or rescue. For some this tactic worked, but for many it exacerbated the problem: as Pugh's studies of the dead boy's clothes showed, once the body was severely cooled, the metabolic rate increased to keep the body temperature at a safe level, and exercise (even just walking) on top of this led to the much earlier onset of fatigue. Sometimes it was better to hunker down, keep still and warm, and wait for rescue.[5]

The interaction between exercise and environmental challenges has been a complication for investigators for the entire period covered by this book. In the nineteenth and early twentieth centuries, Bert, Mosso, Longstaff, and others debated the relative role of fatigue in producing mountain sickness; from the interwar period through the middle of the twentieth century, physiologists increasingly thought that physical work was necessary to develop heat acclimatization in the White body; and throughout the twentieth century, researchers in the Antarctic recognized that whether explorers were man-hauling or riding snowmobiles made an extraordinary difference to their thermal experiences. Fatigue was, in many cases, the difference between the laboratory and the field: even the champions of barometric chamber studies recognized that they could not accurately re-create the experience of months of trekking to Base Camp at Everest. Fatigue was also a complicating factor in fieldwork: exhausted, hypoxic explorers could not do basic arithmetic, fought with colleagues, broke equipment, and made unpredictably bad decisions that threatened their survival.

Studies of fatigue, and related studies of stress, have their own histories in the twentieth century.[6] The phenomenon of "befuddlement" at low P_{O_2} may have first appeared in accounts from nineteenth-century balloonists and from barometric chambers, but it was rapidly taken up as a research problem by military scientists, particularly those involved in early aviation. Physiologists who worked across civilian mountaineering and military aviation used and refined standardized fatigue and mental fatigue tests (such as card sorting) that were later used by extreme physiologists in the Himalaya and Andes. While many extreme physiologists remained interested in discretely measurable bodily fatigue (hence their endless calorie calculations and metabolic studies), those in the field increasingly recognized that, perhaps particularly in challenging environments, fatigue had a strong psychological component. Directly mirroring the findings in stress and fatigue science elsewhere,[7] work on mountains and around the poles showed that a mass of influences—temperature, nutrition, mood, underlying infection, confidence—had significant effects on the work that any specific body could do in any specific situation. In turn, extreme physiologists suggested that their psychological studies and observations would have value in choosing military personnel for stressful situations (for example, submarine service) and, later, that they could be used to select trainee astronauts.[8]

While the complexities of fatigue had obvious consequences for day-to-day life, for diet, for labor theory, even for space exploration, they were literally life-or-death questions for explorers. Fatigue is at the heart of the conclusions Pugh drew from his corpse-clothing experiments to address one of the most important decisions an explorer must make when things go wrong: wait for rescue, or walk to safety? The precise context of the emergency will clearly have an impact on the "right" decision. On Everest, or other high-altitude mountains, for example, the decision is often whether one risks one's own life to evacuate a colleague or ensures one's own survival by leaving that colleague to die. But in the context of the relatively populated and accessible territory of British wildscapes, "seek shelter and wait" was Pugh's recommendation, which, although uncontroversial now, was contrary to much existing advice in the 1960s.

Wilderness and exploration medicine were nascent fields in the middle of the twentieth century (and are medical disciplines that still await serious historical scrutiny). Consequently, much of the eventual consensus on advice and theory relating to those disciplines emerged first from the physiology of extremes in the first half of the twentieth century, particularly from figures such as Pugh, who wrote and researched broadly in this area. Pugh's work on cold immersion led to advice to the "general population" that directly contradicted the existing "scientific" advice for those lost at sea (including military personnel). In 1950, E. M. Glaser, one of Pugh's colleagues at the MRC, had published a letter in *Nature* based on cold bath experiments that logically implied that those who found themselves in cold water ought to swim in order to generate metabolic heat to keep the body warm.[9] A debate ensued, and Pugh flatly contracted this suggestion: taking fatigue seriously meant a new emphasis on staying calm and still and using minimum body energy to float and wait for rescue. This "stay calm" recommendation remains the core of immersion survival: finding ways to deal with the panic reaction that causes gasping breath and inefficient body motion, all of which increase the chance of drowning or hypothermia. (Although, of course, the precise emergency situation still affects the "best" decision—as Pugh's boss Edholm wrote in 1955, perhaps trying to mediate between his two employees, that it might depend on the physique of the swimmer, but "the evidence on this point is not yet adequate.")[10]

This "swim or float" debate is yet another example of the limitations of laboratory models when applied to complicated and often unpredictable real-world scenarios. This book has given repeated examples of idealized mathematical models, and even carefully designed equipment, that failed when faced with Arctic cold or Himalayan transport systems, or even the earth's uneven atmospheric depth. One of the biggest challenges of research into body responses to extreme environments is precisely that it is almost impossible to conduct effective controlled experiments. Pugh's cold chamber research was the closest he could get to a re-creation of the conditions of a fatal hill walk—hence the cold showers, the ergometer, the original clothing—but even here he is clear that this setting differs very significantly from *actual* hill walking and that any findings that emerge from it can only be indicative of the real-world experience. This is one reason why anecdotes and experience remained such a sought-after data source in extreme physiology—hence Pugh's list of twenty-three case studies that he carefully analyzed to try to find commonalities and make suggestions for future preventative action. Here is where the real-world experiences of explorers and athletes proved their crucial usefulness for scientists: they were willing to run 20 miles, to swim in cold water while "wearing" rectal thermometers, to expose themselves to extremes of heat and cold, to climb to high altitude with and without oxygen. As I have detailed elsewhere, Pugh and researchers like him made extensive use of elite athletes in the twentieth century for precisely this reason: in the case of cold immersion, much of Pugh's important data came from the cooperation of English Channel crossers and other elite cold-water swimmers.[11]

The traditions and practices of exploration made possible science that could not have been conducted in any other context, partly for stolidly logistical reasons, but partly because explorers donated their embodied experiences, their physical capacities, and in some cases their body fluids and tissues to experimental work. Their generosity should provoke us to think about the issue of human agency when we consider experimental practices that were frequently dangerous and sometimes fatal. Immersion hypothermia, studied by Pugh using athletes and the dead, links us immediately to people whose agency was disregarded in the most egregious and brutal ways. In the late 1930s and early 1940s, researchers at the Harvard Fatigue Laboratory made studies on "mental patients" who had been diagnosed with schizophrenia and were being treated—against their will in many cases—with intense forms of "cold therapy."[12] There are even more disempowered human subjects: Edholm and Burton's seminal *Man in a Cold Environment* (1955) refers to the Nazi hypothermia experiments—which involved the deliberate murder of prisoners by cold immersion—as "horrible," and yet discusses their results.[13] In turn, this work is cited by Edholm and his colleague A. L. Bacharach in the first book on exploration medicine published in English (1965).[14] It was Luft's apparent knowledge of these experiments that eventually undermined his reputation after he managed to create a new life in the US. Since 1948 scientists and ethicists have debated whether it is acceptable to benefit from Nazi research when it involved the torture and murder of human subjects. Roughly speaking, three schools of thought exist: first, that the research has been done, and that failing to apply the knowledge would mean the subjects suffered and died for no reason; second, that the research was "bad science," poorly designed and unreliable, and should be discarded for that reason; and third, that regardless of the usefulness or accuracy of the results, it is simply unethical to use it.[15] All three of these options were, at one time or another, chosen by the extreme physiologists who have peopled the previous chapters.

There is nothing as egregiously exploitative as the atrocities of the Second World War in the studies and expeditions discussed here. Nonetheless, there are questions to be asked about the participants in experiments and in the physiological work done during expeditions. There have clearly been practices used that would not pass today's ethical standards—for example, the 1953 British Mount Everest expedition team's decision to test the benefits of amphetamines by giving them to a couple of Sherpa porters working on the deadly Khumbu Icefall[16]—and cases in which existing rules have been bent or waived for the benefit of extreme physiology research—for example, in the funding of Everest expeditions (more on this below). There are also situations in which subjects' willingness or unwillingness to participate and ability to give or withhold their consent are more complex.

Along for the Ride

The majority of human subjects in the extreme physiology studies and experiments considered in this book were explorers, mountaineers, or athletes—often also scientists and physiologists themselves, since much of extreme physiology is self-experimentation. Chris Pizzo's sampling of his own alveolar air on the summit of Everest is the seminal example, as on that expedition his official role was "scientist-climber." Self-experimentation has a long and robust history in physiology, and for extreme physiology it was usually practiced for either a pragmatic or a persuasive reason. Pragmatically, self-experimentation was easy and convenient, and it guaranteed a cooperative and informed human subject. Self-experimentation was crucial to the earliest work in altitude physiology, beginning with Bert's studies of his own body (as well as those of accompanying animals) in barometric chambers in the 1870s. Likewise, Haldane, Douglas, Mosso, Zuntz, and others became subjects of their own experiments not only on mountainsides, but also in chambers.[17]

Self-experimentation (or at least self-sampling) was also a technique used to persuade other human guinea pigs to take part in experiments willingly. Several experimenters noted that testing themselves seemed only "fair." Ove Wilson, medical officer to the Norwegian-British-Swedish Antarctic Expedition (1949–52), who took fairly extensive blood samples through the duration of the expedition, came up with a scheme to encourage and sustain participation:

> For psychological reasons I did not take the blood samples on myself, but let them be done by any of my comrades who liked to try their skill. In this way some of them took a "cruel revenge" on their torturer, often inspired by the spectators' delighted remarks.[18]

(As a side note, this is also an example of the ways in which extreme physiologists and participants in expeditions worked of necessity across disciplines and learned new skills as a result; as Wilson adds, "Most of my comrades attained considerable skill in blood-letting during those two years.") Similar approaches were tried on mountains. During the 1936 British Mount Everest expedition, the medical officer, Charles Warren, made a series of medical and physiological investigations, some of which were quite intense, including the detection of changes in gastric acidity at different altitudes using a test meal:

> The most unpleasant part of the performance of this test is persuading your victim to swallow the stomach tube. It was only fair therefore that I should set the example by doing the first test-meal on myself.[19]

For explorers and athletes, time spent in a barometric chamber, or a climate chamber, could be considered as much training as experimentation. Several members of the various British Mount Everest expeditions spent time in barometric chambers and wind tunnels, in part to "experiment" with equipment to test its suitability, but also in part to test themselves, as, for example, by practicing walking in adverse conditions wearing oxygen masks.[20] For some participants, particularly athletes, physiology experiments and studies had such immediate and obvious utility to them that they actually sought out opportunities to be tested. Take as an example the relationship between the British middle-distance runner Martin Hyman and Pugh through the late 1960s. Hyman was one of the runners chosen to go

9. Self-experimentation in progress at Camp IV on Everest, 1953. John Hunt (*left*) assists L. G. C. E. Pugh (*right*) in taking a sample of his own alveolar air. (Photograph by Alfred Gregory. © Royal Geographical Society, London, UK.)

on the British Olympic Association/Sports Council/MRC–sponsored field trip to Mexico City to investigate the effects of mid-altitude on athletic performance.[21] After the trip Hyman and Pugh remained in close contact, with Pugh sending Hyman offprints of his articles, and Hyman offering his own (and others') bodies as experimental objects—"With regard to further work I should be pleased to help in any way I can"—and suggesting research projects:

> I should be very interested to see a well controlled experiment to find out whether there is a significant difference in O_2 consumption during a short period of work when it is and is not preceded by warm-up.[22]

Hyman clearly saw the value of Pugh's research to his own training and performance, and later, in the early 1970s, began to seek Pugh's assistance in designing fitness training and testing regimes for orienteering, a sport in which he had begun to coach.

So it is important to note the agency of many of the human participants in extreme physiology experiments. Athletes in particular seemed interested in receiving feedback on their performance and physical attributes—bear in mind that through most of the twentieth century it was not possible for most runners or climbers or swimmers to monitor their own heart rate, much less their metabolic processes, hormone levels, and so forth. Similarly, many participants in extreme physiology experiments were themselves trained scientific professionals, and while they might not be experts in biomedicine, they were aware of the traditions and expectations of scientific practice and of experimental processes. Such experiments therefore blur the boundary between expert and subject, and yet they are possibly the most common form of human experimental practice—where "being a guinea pig" is a process of fair exchange, tit for tat, or reciprocal favor granting (offering one's own body, or sometimes those of students, laboratory assistants, etc.). The example discussed in most detail in this book is Allan Rogers's IMPing experiments in Antarctica in the late 1950s, in which the geologist Geoff Pratt in particular (though there were dozens of other subjects) put up with extremely inconvenient and intrusive experimental equipment for the sake of another's research program, provided detailed feedback on the conduct of the experiment despite not being a physiologist, and in turn "recruited" Rogers as a "geologists' mate" to help with Pratt's own research. Those who volunteered to take part in experiments might be doing so to produce either general or specific knowledge of relevance to themselves (training tips, clothing advice), but they might also be trading their bodily labor in return for assistance with other tasks—cooking, reading meteorological equipment, or hauling a sled.

Willingness to take part in scientific studies in extreme conditions might also be motivated by money. Although explorers were almost never paid directly, other participants were sometimes offered financial rewards. In an article published in 1961, the MRC researcher W. R. Keatinge reported that he had failed to recruit as many naval personnel for his cold chamber experiments as he had hoped, and had therefore offered the men both increased pay and extra leave as an "inducement to volunteer."[23] Even absent such inducements, explorers could still have essentially financial motives to take part in extreme physiology experiments—not least because such work could be the difference between a fundable expedition and one that would bankrupt all the participants. As individuals, scientists had access to funding sources that explorers could not otherwise exploit. National scientific organizations, from the Royal Society of London to the National Institutes of Health to the Australasian Association for the Advancement of Science, also contributed to expedition expenses in the twentieth century. Science-minded employers, such as university departments, the MRC, and NASA, were sometimes willing to continue paying wages to or cover travel costs for employees who got places on expeditions. In 1959, as Hillary and Pugh began to plan what would become the Silver Hut expedition and Hillary was worrying about funding, Pugh pointed out that "one might get enough money from scientific sources alone for a scientific party to accompany a mountaineering party."[24] The science, not unlike the yeti hunt, was a way to raise money for what probably interested Hillary more: the expedition. This reliance on science was not a one-off—as Phillip Clements has shown, tapping into Cold War priorities and making (as it turns out, undeliverable) promises about scientific research was the only way the participants in the American Mount Everest Expedition (AMEE) of 1963 could imagine getting the funding they needed for a high-altitude Himalayan adventure.[25] The fact that much Antarctic exploration was justified or funded by scientific work is a truism of polar studies—not least because access to the continent itself has so often been predicated on scientific activity (as nominally opposed to territorial or military interests).[26]

Such funding does not come without ethical considerations. The first specifically biomedical expedition to Everest was the American Medical Research Expedition to Everest in 1981 (AMREE—not to be confused with the 1963 AMEE, which included a limited amount of physiology in a rather more diverse and ambitious scientific program). In their appeals for funding, the AMREE organizers—principally the physiologist John West—highlighted not only the failings of barometric chamber experiments, but also the problematic ethics of confining athletic young men to chambers for weeks, if not months, at a time. And yet, taken in the abstract, the choice of AMREE's multiple funders (including the National Institutes of Health, the American Lung Association, the US Army Medical Research and Development Command, and the National Science Foundation) to pay for an expedition to Everest is extraordinary under the usual ethical conditions—because the mountain is deadly.[27] The risk changes depending on the exact criteria used (e.g., deaths per summit, deaths per person on the mountain, Sherpa vs. Sahib deaths), but by the early 1980s it was not insignificant: probably around one in fifteen summiters died as a result of their expeditions. It is difficult to imagine that another experimental protocol containing an acknowledged risk of 5%–10% *mortality* for healthy, active, fit young men would ever receive ethical board clearance, let alone funding. The obvious exception is space travel; clearly, some fieldwork, particularly that in extreme environments (both on- and off-world), allows different understandings of risk and reward.

Unlike the risks of nearly any other imaginable physiological study, the risks of extreme physiology are generally not a result of biomedical processes—blood tests, being attached to a Douglas bag, or even the insertion of a gastric tube—but rather the result of the environment in which those tests are taking place. These risks were not limited to cold, altitude, and rough terrain, but often involved the infrastructure of exploration itself: some of the most potentially harmful scenarios arose as a result of carbon monoxide poisoning in underventilated huts (or as a result of human failures in the process of setting up and handling portable stoves).[28] In a counterintuitive way, the significant risks of exploration can result in a diminution of the ethical burden of human experimentation—the subjects are young, fit men choosing freely to engage in an activity with a significant risk of mortality and morbidity. Blood tests and sessions riding on an ergometer seem like no burden at all when balanced against the odds of injury or death the participants have already freely chosen. This is not to argue that the medical procedures were universally safe—on the 1981 AMREE, one climber developed a rash and a fever in an allergic response to an injection containing albumin, an outcome that the organizers had not considered as a risk to mitigate.[29] The willingness of experimental subjects to participate in risky biomedical work should not make us oblivious to the fact that there can be a level of coercion operating in these situations: it is clear that the draw of Everest or Antarctica is such that there is competition for access, and if the price of access is agreeing to bodily surveillance, taking part in experiments, or promising to fill in record cards, that is a price many are willing to pay. In other words, *the expedition is the payment* for participation.

Participants therefore had many motivations for taking part in extreme physiology research, few of which were strictly altruistic, in that many expected some benefit from taking part. Even if they were not paid in cash and vacations, they were often given something less tangible as a reward for their role in the experiments. This did not even have to be something as glamorous as a trip to Everest or the South Pole—any "big mountain" or remote location might be a draw. In the late 1960s and '70s, Charles Houston was involved with the Mount Logan High Altitude Physiology Study, a long-term series of altitude physiology experiments based at a research institute on Mount Logan in Yukon, near the border with Alaska. At first the experiments were conducted on Canadian military personnel, but although "they were fit, strong, tough, independent and obeyed orders to the letter . . . they weren't particularly interested in what [the experimenters] were trying to achieve."[30] Consequently, the researchers decided to try to recruit from the "many young people [who] were interested not only in the chance to be on a big mountain but also in the work itself."[31] Each year fourteen to sixteen volunteers, most aged between twenty and twenty-nine and mostly medical or physiology students, were selected to be "subjects, collaborators, and support personnel" (it should be noted that this project was unusual in that it accepted female applicants from the start).

Similarly, expeditions are of necessity tightly ordered activities (at least, ideally so), involving hierarchies, systems of leadership, and routines. Certainly on the smaller-scale mountain expeditions, and on most of the circumpolar expeditions, there is an expectation of willing participation beyond one's prescribed expeditionary role; part of being a "reliable chap" is not being a "jobsworth," that is, being someone who is willing to take up slack, help other team members with their tasks, and participate whenever and wherever labor is needed. That said, there is some evidence that, at least in a joking way, explorers would occasionally try to avoid taking part in biomedical experiments. John Hunt wrote of Pugh's maximum work test as a "fearful ordeal" and added that "there had been sacrifices to science which I was glad to avoid. . . . We hurried forward to catch up the others lest we should be tested in our turn."[32] (Note, too, that this test is another case in which self-experimentation was deemed "fair" practice: "It was satisfactory to learn that Griff [Pugh] . . . had not spared himself the tortures which he inflicted on his guinea-pigs"—see fig. 9 above.)[33] But, overall, subject compliance is an issue about which extreme physiologists rarely complained, and the value of disciplined, routinized team players is also clear, not least from studies like Rogers's acclimatization study on the TAE: sixteen men were expected to make daily records of their clothing, mood, sleep and work patterns, and so on, for a period of more than a year, and yet, as Rogers put it:

> Despite the arduous conditions and extremely long hours, the cards were filled in regularly and faithfully by all the men. Only at the end of the crossing is one set of cards really short of data.[34]

Many explorer-scientists showed an acute awareness of the intrusive nature of their work—at least when it was being practiced on White and Western bodies—which is often evidenced by effusive and generous thanks in the acknowledgments to articles. These spaces are also useful for the historian in that they often also acknowledge the trauma or ethical challenges of the work being conducted. Tests that can sound relatively innocuous at first reading can turn out to be extraordinarily stressful: Charles Eagen's 1963 paper on "finger cooling" used mountaineers and US Air Force personnel, as well as members of various North American indigenous populations, to test the relationship between fitness and cold resistance, and while "finger cooling" might not sound extreme, Eagen writes, "I thank the test subjects for their co-operation and especially for their forbearance during the tests which caused *severe cold pain* [my emphasis]."[35] Alternatively, acknowledgments might slightly euphemize medical interventions, as in one South African study of British Antarctic Survey workers' adipose fat, published in 1967, which specifically thanks the human guinea pigs for putting up with "a somewhat distasteful operation . . . with unflagging co-operation."[36] This "distasteful operation" was in fact a regular fat biopsy from the buttock, and the cooperative BAS workers had fat "punched" out of them up to eleven times in one year.

Clearly, some physiologists show more awareness of their reliance on cheerful human volunteers than others, and this awareness is not dependent on the time period—if anything, the early physiologists, working outside a culture of regular medical investigations in extreme environments, sometime appear even more careful about their subjects than those at the end of the century. During what was probably the first major physiological work done on any expedition to an extreme environment (preceding Pugh's work on Cho Oyu by a year), H. E. Lewis wrote of the British North Greenland Expedition that the "physiological tests were not pleasant for the subject," not just because of physical discomfort due to blood studies or being woken early from sleep, but also *psychologically*: "The subject may well have felt conscious of his dirty body and unwashed clothes, which were of necessity a feature of the expedition."[37] Lewis and the medical officer, J. P. Masterton, subjected themselves to the same studies, just as Pugh did, but Lewis also noted that

> we were dealing with a group of educated men, whose scientific attitude required a logical explanation for the medical research programme. Once satisfied, they cooperated actively.[38]

This view contrasts rather sharply with the experience of the IBEA twenty-five years later, in which one major finding was that "few of the scientists had any real understanding of group dynamics or of procedures by which subjects could be humanely and carefully treated."[39]

Explorers, like athletes, are not only (usually) willing to put themselves in extraordinary situations for the sake of science, but are also promising guinea pigs in terms of compliance and reliability. In addition, these human subjects are often "trained" in ways that prepare them to participate in the studies: the closed and self-regulating nature of extreme physiology and expeditions, as laid out in chapter 3, means that explorers and sporting participants were likely to have taken part in several physiological experiments, or at least to have experienced practices such as fitness testing, which in some degree would familiarize them with the investigations and equipment favored by physiologists. Some of this equipment could be difficult to work with. The IMP is the clearest example here: even after many days of wearing the equipment, Pratt still struggled to do a "normal" day's work, a finding that had implications for the conclusions of IMPing studies elsewhere that had used less self-aware subjects, or subjects in less "natural" conditions who might not notice a change in their routine behavior. (This was also an experimental method in which psychology had to be considered—in a later IMPing study, the MRC physiologist J. R. Brotherhood noted wide variation in the energy expended doing domestic chores in Antarctica, which he put down not to the inconvenience of the IMP itself, but rather to "the individual's attitude to the job in hand.")[40]

Subjects' useful experience did not have to come specifically from previous experimental work, but might also come from sociocultural and sporting practices. Although the Silver Hut team wanted to expand their studies to the Sherpa as well as White climbers, they found this difficult, in part because their high-altitude porters did not know how to ride a bicycle.[41] As the ergometer (fixed bicycle) was the standard piece of equipment used to simulate exercise, measurements from Sherpa porters who took part could not be reliably compared with those from more experienced White cyclists due to the difference between the two groups in pedaling efficiency. During metabolic and exercise studies at altitude (which involved "Peruvian indians" as well as Sherpa), Italian teams in the 1970s deliberately swapped out the ergometer for uphill running tests, as "several subjects, particularly the Sherpas, were unable to pedal."[42] White researchers also thought they faced further challenges in studying indigenous peoples because, as the British researcher Roy Shepherd wrote in the 1970s, "there are difficulties in motivating primitive and non-competitive people to perform an all-out effort."[43] Motivation is not, obviously, a race-limited characteristic, and other physiologists made similar complaints about White participants through the twentieth century. Indeed, the British physiologist (and Nobel Prize winner) A. V. Hill justified his focus on athletes as experimental subjects in part precisely because they worked "to the utmost" of their powers and were therefore more motivated than, say, industrial workers or student volunteers. He added that they were generally healthy adults that could "be experimented on without danger" and were able and willing to "repeat their performances."[44] It was exactly this combination of features—willingness and ability—that made explorers and mountaineers such attractive subjects of study for twentieth-century physiologists.

Following Orders

There was another group of (mostly) men who, it was thought, could be relied on to work to limits and follow strict instructions: military personnel. There were deep connections between military activities and exploration science throughout the twentieth century. These connections could be intellectual, as with the recruitment of high-altitude and mountaineering specialists such as Barcroft, Haldane, or Luft into military research on gas poisoning or aviation, and the reverse pattern of recruiting military and ex-military personnel for work on survival science, space programs, and so on. They could also be logistical, from the use of navy vehicles as transport to Antarctica to the military-style "assault" planning of mountain expeditions that became more popular in the period immediately after the Second World War, epitomized by John Hunt's organization of the 1953 British Mount Everest expedition. Or they could be material: we have seen clear interchanges of respiratory equipment between the RAF and high-altitude expeditions, and many expedition food systems started life (and were first tested) as military ration packs. And they could be personal: through the nineteenth century the association of exploration and colonialism meant that many travelers were "on Her/His Majesty's service," and conscription and global wars meant that many explorers in the middle of the twentieth century had had experience in the armed forces. For some (ex-)military personnel, travel was the reason they became interested in exploration. Pugh, for example, traveled to Greece and India and across the Middle East in the Royal Army Medical Corps, as well as training ski troops and working for the Mountain Warfare Training Centre in Lebanon.[45] All this meant that many of the bodies used in, and making use of, extreme physiology research were current or ex-military employees.

Some of this research does appear coercive—to physiologists involved in the discipline as well as to historians writing in a twenty-first-century ethical landscape. As we saw in chapter 2, John West, in a letter written to justify the expense of a 1980s Everest expedition, pointed out that the only known long-term barometric chamber experiment, which had taken place nearly forty years previously, "was done during wartime and the naval recruits *had little option but to agree* [my emphasis]."[46] West was framing the problems of barometric chamber studies as fundamentally part of the problem of mapping artificial laboratory situations to real-world field studies and highlighting the role of "real" fatigue (rather than artificial exercise on stationary bikes in confined spaces) as part of the process of altitude acclimatization. But at the same time, his aside about military personnel has some real weight: the chamber experiments he is referring to are the Operation Everest studies of Charles Houston and Richard L. Riley, and as we also saw in chapter 2, the conditions for the subjects were harsh.

At around the same time, the US military was using its troops to study survival in desert environments—including a plan to "train men to do without water," until physiological studies suggested that making men march without sufficient water was a "patent error."[47] The earliest substantial book in English on desert survival—*Physiology of Man in the Desert*, published in 1947—contains multiple examples of experiments on American recruits, who sweated to near exhaustion and collapse in hot rooms, in the swamps of Florida, and at sea without water in small boats.[48] All this is not to suggest that military personnel had no agency in choosing whether or not to participate in experiments, nor that they derived no benefit (or interest, or pleasure) from taking part; at least sometimes they were given extra pay and leave, as was the case for the naval personnel taking part in the MRC-funded cold study cited above. As with explorers, in many instances the challenging aspect of an extreme physiology experiment—at least when conducted in the field—was something the soldiers, sailors, or aviators were going to do anyway; the desert march was going to happen whether or not the physiologist asked you to swallow a pill thermometer; the stint in Antarctica was simply a point on a military rota of duties, regardless of whether or not you filled in your nutrition and clothing form at the end of the day.

Local Bodies

Despite any difficulties with cycling proficiency, extreme physiology has made extensive use of the bodies of people who live permanently in non-temperate and high-altitude environments. Probably the most scrutinized have been the mid- and high-altitude residents of South America, who were studied first by European and North American scientists, from Viault in Peru in the late nineteenth century to the members of the International High-Altitude Expedition to Chile in 1935. The same populations then became the subject of autochthonous South American altitude physiology research with the founding of the Instituto de Biología y Patología Andina in 1934. As outlined in chapter 5, the fact that it was Andean populations at altitude, and not Himalayan ones, that were initially studied probably had a significant effect on the study of respiration and circulation at altitude, as the two populations present different acclimatization and adaptive mechanisms. Meanwhile, circumpolar peoples were frequently the subjects of anthropological, physiological, and genetic study (particularly, for the latter, during the IBP). This category obviously covers an extremely diverse range of populations across North America and Eurasia, and early hereditary and blood group studies emphasized that diversity, so, in contrast to altitude, it was not an intellectual leap to expect varying adaptations to cold. Researchers were largely disappointed in these investigations, though, as their studies—whether physiological or genetic—repeatedly showed no or little evidence for anatomical or hereditary adaptation to cold. Cold survival seemed much more a matter of habit and technology; essentially all bodies could become "Rodahl's Eskimo" if they adapted their dietary and working patterns and used the most efficient means of protection from the weather, whether clothing or shelter.

Theories of acclimatization and adaptation to cold and to heat were essentially opposites: a strong assumption remained of hereditary adaptation to hot dry and to hot humid climates, evidenced not least by skin color (at least after midcentury).[49] There were researchers who tried to argue for alternative narratives, such as Macfarlane, who repeatedly tried to demonstrate in his work with Australian Aboriginal peoples that their "hereditary" adaptation was for hot, humid climates, and that their ability to survive in the dry heat (and, overnight, dry cold) of Australia was thanks to cultural and technological adaptations—but this association of culture and technology with adaptations to hot climates remained a minority approach right through the twentieth century. This research bias, that acclimatization to cold was technological and adaptation to heat was biological, usefully reinforced an existing assumption of superiority (in terms of culture and civilization, if not biology) for temperate populations. The "out of Africa" hypothesis of human evolution was read as indicating that populations in hot climates were *literally* closer than other populations, in an evolutionary sense, to primitive human ancestors. This assumption was one of the driving hypotheses of the Human Adaptability theme of the IBP, which sought out "untouched" populations for genetic study, hoping that they would provide a "baseline" for human genetics—a coded way of saying "primitive" or "closer to primitive ancestors."[50] The interactions among studies of indigenous non-temperate populations, racial science, and evolutionary theories is the proper topic of another book, but in the work of extreme physiologists the basic features of this relationship can be clearly seen.

Any ethical critique of these practices has to be alert to the agency of the indigenous peoples involved in research. It would be a repeat of the paternalism of some explorers and scientists to assume that these relationships are necessarily exploitative—to frame indigenous peoples as unable to make informed choices about participation, or to assume they were unable to find ways to benefit and profit from experiments or scientific expeditions. In particular, Sherry Ortner's detailed anthropological and ethnographic work among the Sherpa has demonstrated how the local communities of the high Himalaya adapted their economic, social, and cultural practices to absorb the sudden influx of money and employment that came with Western high-altitude mountaineering. It is clear that "Sherpa" and "Sahib" sometimes occupied almost parallel worlds, in which Sherpa activities were interpreted through a Western lens and in some cases actively misunderstood.[51] For example, the Western tendency to celebrate Sherpa "loyalty" (which itself highlights a colonial assumption that indigenous people lack this quality) failed to recognize the crucial role of mentorship in Sherpa culture. Demands— enforced by withholding labor on the mountain—for better equipment, particularly shoes and tents, were variously interpreted as a cultural matter of respect and hierarchy within a group of porters; a matter of fairness where all climbers deserved the same technological support; or an indication of the increasingly "commercial" attitudes of Sherpa, who now wanted "Western" technology and high-value items. Whatever the interpretations, they demonstrate two things: that there *were* inequalities in the provision of survival technologies on expeditions, and that indigenous peoples had the mechanisms and the inclination to object to such inequalities.

When it comes to the ethics of consent and risk, it is clear that financial compensation and social status acted as incentives to Sherpa assisting with high-altitude climbs, some of which ended in Sherpa fatalities. But, of course, all participants had their motivations: as we saw in the section above, many participants were effectively "paid" by boosts to their careers or access to unique sporting opportunities. It is pertinent, however, to ask who benefits from the research that indigenous people supported. Certainly for the first half of the twentieth century, a fairly straightforward Marxist or labor-rights critique could be applied to some extreme physiology research. It was, after all, miners who were being studied in Cerro de Pasco—manual laborers working long hours, in difficult and dangerous conditions, for comparatively low pay. Access to the workers was granted by their employers, and we can certainly see this employer-researcher relationship as being part of a broader set of relationships that tied research, efficiency, profit margins, and "scientific management" together throughout industry and science in the first three or four decades of the twentieth century.[52] The Harvard Fatigue Laboratory, which sponsored Ancel Keys's trips to South America in the 1930s, was, after all, a joint venture between Harvard Medical School and Harvard Business School, whose remit was supposed to be the scientific study of labor. While some of this work was about health and safety—how hot a factory could get without harming its workers; how long dam builders could work in the sun without increasing sick leave—it was also about discovering the limits to which the human body, as a piece of machinery, could be pushed in the pursuit of profit and efficiency.[53]

Therefore, without wanting to undermine the agency of those who took part in extreme physiology research, we must still ask, "Who benefits?" While this research led, at least in a roundabout way, to health and safety rules and guidelines about operating temperatures, length of shifts, and so on, where organized labor had a significant voice, it objected to efficiency mechanisms that sought to push people to the limits of production and performance.[54] It is not clear that the specific research at Cerro de Pasco or elsewhere had any direct positive influence on the health or working lives of the indigenous people who took part in the studies. Likewise, it is extremely difficult to draw a clear line from high-altitude or circumpolar research to direct benefits for indigenous people in those regions. It is particularly obvious that while a great deal of effort went into technological support for Western explorers, the needs of guides and assistants were a distant second: for example, even in the 1960s the face masks for Western oxygen equipment were still being adapted on-site in an ad hoc way to suit the "flat noses" of the Sherpa—that is, after forty years of climbing with oxygen and with Sherpa, no Western team had designed a mask that suited their facial physiognomy.[55] This neglect went beyond ignorance and into essentializing racism. On the 1971 International Himalayan Expedition, an expedition to Everest that celebrated its collaborative and global climbing credentials,[56] organizers took two kinds of face masks from the US Air Force, marked "Caucasian" and "Oriental," assuming that the latter would work for their "oriental" Sherpa companions. Based on Vietnamese facial physiognomy, the "Oriental" masks were, if anything, a worse fit than the traditional RAF-modified masks used by previous American teams.[57] Likewise, while the boots of Western climbers were usually tailored to their feet, boots for Sherpa were bought in bulk, usually in a set of standard sizes or based on some average measurements, to be adjusted by the wearer on the mountain itself. These acquisitions did not always go according to plan; for example, as Hugh Ruttledge, leader of the British Mount Everest reconnaissance expedition of 1935, writes:

> On March 12 we made an issue of kit to the porters, and had a great fright when it was found that large numbers of the marching boots supplied for them from Cawnpore were too small. How this had happened I do not know, for Morris had taken the trouble at Darjeeling to measure the feet of several typical Sherpas and Bhutias, and to send measurements and outline to Messrs. Cooper Allen at Cawnpore. I was obliged to . . . [order] 110 fresh pairs.[58]

While this problem may have resulted from a mistake by the shoemaker, it would not have occurred had organizers not assumed that one Sherpa foot was, after all, much like another.

Some circumpolar research in the Far North was translated into public health interventions (particularly in the case of Alaska), and in both Arctic and high-altitude investigations, explorers often offered medical services, both to their own porters and guides and more generally to the villages through which they passed. Much of Kåre Rodahl's work among the "Eskimo" was predicated on an exchange of medical services in return for physiological and anthropometric measurements of local people, and accounts of most European Himalayan expeditions explicitly mention that the team's medical officer was in high demand to treat illness and injury on the treks to base camp. The use of medical treatment as a way to "win hearts and minds" is a familiar practice in colonial and postcolonial history, whether the doctors were selling Christianity or Western medicine. This practice, too, has problematic ethical aspects, particularly when the boundary between experimentation or study and treatment is unclear (or treatment is provided only on the condition that the person submits to study): the most cited example in the twentieth century is the Tuskegee Study, in which participants were told that purely surveillance interventions (e.g., a lumbar puncture) were actually treatments for their syphilis. Nothing that egregious appears in the histories discussed in this book, although the use of amphetamines on the Khumbu Icefall in 1953 provokes ethical questions that, now that all the participants have died, may never be answered.

Silent Witnesses

There are participants in extreme physiology research who can neither consent to, nor benefit from, its results because researchers often learn from the dead. This chapter opened with the example of Pugh's research into hypothermia in Britain's wilderness spaces, which demonstrates two ways in which the dead were used—as anecdotes and as experimental objects. The dead could be used in case studies as examples and lessons based on the knowledge that could be gathered about the circumstances and nature of their deaths; this use could often be fairly informal, even based on hearsay (and should call to mind the very start of this book and the mass of anecdotal evidence that Bert corralled to provide the starting point for his research into altitude). It is more unusual, and perhaps more ethically complex, for the dead to be used in a process of active experimentation, such as Pugh's experiment with clothing taken from a victim of the Four Inns Walk. There is also an intermediate situation, in which taking data from people in life-or-death situations who subsequently die effectively turns death into an experiment. Clinical measurements taken from the recovered bodies of people lost in the desert, such as rectal temperatures, are used to provide information about the effects of extreme heat and dehydration on human bodies—effectively a "natural experiment" examining a process that cannot ethically be studied elsewhere (so, too, are the anecdotal stories of those who have survived such traumas: one such story cited in scientific papers is that of Mauro Prosperi, a runner who got lost during the 1994 Marathon des Sables and survived by drinking his own urine and eating bats).[59]

Of course, fatal experiments *have* taken place, and they may well continue—as discussed above, data from the Dachau hypothermia experiments was regularly cited in works on extreme physiology through the middle of the twentieth century, albeit usually with a disclaimer about the unethical source of the information. The essence of extreme physiology is, after all, a fascination with the limits of the human body, both in terms of its tolerances and its performance. This fascination is not limited to short-term acclimatization or exposures—the Human Adaptability theme of the IBP specifically sought out populations that lived at environmental extremes in the hopes that they would reveal the "limits" of human biological evolution—the most diverse and varied racial forms of humanity. When it comes to short-term adaptation and acute exposure, however, the accounts of extreme physiology experiments clearly demonstrate that some human subjects were indeed pushed to the limits of performance: athletic participants collapsed, exhausted, vomiting, shivering, while inside them a radiothermometer sent back readings of their internal temperatures; heat chamber experiments resulted, if not in collapse, then in subjects whose

movements were not well co-ordinated. They stumbled. They paid little attention to the observers, occasionally threw their arms about, and eventually refused, or were unable to go on . . . irritability was frequent, and outbursts of bad temper or emotional weeping were not uncommon, when men were nearing the limit of their endurance.[60]

To properly define a limit of human endurance, it was necessary to examine not only those who stopped just before it, but also those who transgressed it and died as a consequence.

When it comes to taking data from dead human bodies, extreme physiology begins to blur into forensic science. The forensics of wilderness medicine is a topic that deserves—but does not yet possess—its own history, as there is an abiding fascination with the fate of "lost" explorers. Throughout the modern period scientific methods have been used to reveal the "truth" of a death or an accident—sometimes in disregard of the oral testimony of witnesses (as in the case of Franklin) and sometimes in complex combination with it (as in squaring the positions of corpses with "last known sightings" on mountains). This work also blurs into the cultural practice of relic hunting. As already discussed, participants in both exploration and expeditionary science have a strong affinity for nostalgia and heritage, deliberately linking themselves to previous explorers and expeditions and feeling tied to them by the limited human pathways available up mountains and across polar spaces. Part of that practice involves the collection of items that can be both relics and scientific "evidence"—Mallory's clothes, for example, became the "evidence" in studies of the adequacy of his 1920s outfit compared with twenty-first-century survival clothing, while Hannelore Schmatz's ice axe was a "miracle" that enabled the first clinical measurement on the summit of Everest.

As this book has demonstrated, one of the complications of extreme physiology and expeditionary science is that it usually takes place within a complex of other practices and motivations: almost no expeditions are "purely" biomedical even in intention, and in reality, sporting goals, military strategy, geographical ambitions, and expressions of national and gender identity, as well as other forms of scientific practice (including yeti hunting), all shape and refigure physiological work. Such complications apply to forensic work, too: particularly in the analysis of failed or fatal expeditions, the need to learn physiological lessons from past deaths is almost inextricable from the desire to allocate or remove blame. In English-language sources, it is Scott's legacy that is the most picked over. The preservation of Antarctic huts and sites allows for a form of "forensic archaeology"; that is, analyses—sometimes done from an armchair by looking at photographs—of routes, rations, even signs of psychological disorder in the arrangement of accommodations. While authors, including explorers like Ranulph Fiennes as well as academics, have used interventions such as nutritional analysis of early twentieth-century rations to explain the failure of Scott's team, their motivation has been in part to recover the heroism of Scott himself as much as to learn nutritional lessons for future expeditions. Such analyses have extended to retrospective diagnosis of scurvy or mental disorder among team members, so perhaps frustratingly for those interested in this line of investigation, the bodies of Scott's team were never recovered or repatriated. The bodies of Oates and Evans remain lost to the continent, while the corpses of Bowers, Wilson, and Scott were re-covered with snow after their discovery in 1912. With the repeated layering of snow and the geophysical mechanisms of the continent working on them, they probably will (or already have) become encased in ice far beneath the snow layer and be moved glacially across the continent before eventually being deposited in the sea.

To be entombed and moved about by meteorological and geological processes is not an uncommon ending to embodied experiences of extreme environments. While the bodies of the dead in hot and temperate zones tend to rather static desiccation or decay, those who die in snow, whether in circumpolar regions or at high altitude, often become part of a remarkably mobile geological process. Some bodies at high altitude have become famous almost as way-markers for the traditional routes up a mountain—that of Hannelore Schmatz remained in place for over a decade, and in an anecdote Sherry Ortner repeats from 1988, "When the snow recedes off [Schmatz] in the springtime, descending climbers leave empty oxygen tanks [around her] or pat her head for good luck."[61] The body referred to as "Green Boots" (probably Tsewang Paljor), who was a victim of the 1996 season on Everest, remained in position for over two decades on the North Face, outside a site known eponymously as Green Boots's Cave, to be viewed by almost every climber taking that route. The removal of bodies from a high mountain is expensive and dangerous—two men died in the attempt to remove Schmatz's remains—but the mountain itself acts to relocate and conceal corpses. Bodies may be swallowed and revealed by glaciers more than once because of changes in local climate or climbing routes.

These mobile bodies can be used as evidence in scientific work: one study of corpses revealed by glaciers in the Alps in 1991 describes four distinct ways to "use" them.[62] First, those whose fates are well known can be sources of evidence of the pathological and mechanical effects of glacier encapsulation, so that, second, causes of death can be confidently established forensically for victims retrieved from the ice whose history was not previously known. Third, and relatedly, the study of recent victims can throw light on ancient specimens, most famously in the case of "Ötzi the Iceman" who had been preserved in mountain ice for around five thousand years. Studies of these bodies provide unique insights into human prehistory. Fourth, and finally, when the times and locations of deaths are known, human bodies also act as "tracers" showing the movement of glacial ice, thereby providing insights into the complex system of water and ice dynamics as well as demonstrating the effects of global climate change and warming. The global interconnectedness of extreme environments is only emphasized by the fact that the 1991 forensic paper was stimulated by an exceptional glacial melt in that year—caused not by global warming per se, but by unusually high deposition of sand from the Sahara on the Tyrolean glaciers, which increased their absorption of solar radiation and thus their rate of melting. The hot, cold, and high places of the earth are intractably connected.

The View from the Summit: Conclusions

Although, as previous chapters have shown, the connections between hot and cold spaces on the earth could be a product of the subjective human imagination as much as any objective meteorological, geophysical, or geographical similarity, their interactions mean that this book has been of necessity, as well as by choice, a global history. The boom in interest in the history of environmental science has provided us with excellent case studies for the examination of the international aspects of scientific practice, and this history of extreme physiology is another case in point. Internationalism was essential for funding, access, logistics, and expertise; at the same time, it was also a burden and a challenge, sometimes in deeply pragmatic ways, as the International Himalayan Expedition (to Everest) of 1971 discovered when its Austrian-made crampons failed to fit properly on its German-designed boots.[63] This internationalism, however, was a predominantly White male affair. While Nazi science (and scientists) could on occasion be unproblematically absorbed into extreme physiology, it proved much harder to involve—on an equal basis—non-Western peoples and women of all races. Racialized assumptions—whether that all high-altitude peoples must share biological features, or that Australian Aboriginal peoples had no technological or cultural practices of value to White Australians—have demonstrably affected, not to say retarded, Western research. Relatedly, this book has opened up a new area of historical investigation: it is clear there is a story to be told about the relationship between theories of short-term acclimatization and theories of long-term adaptation. While this book has focused on acclimatization, evolutionary theory has still intruded (particularly in relation to altitude), and racial science has been a repeated theme (particularly the robustness of the assumption that cold-climate populations are culturally superior). From the use of "primitive" peoples by the IBP to the use of altitude science in nationalist political polemics, the preceding chapters have hinted at a link between midcentury evolutionary theories and whole-body physiology.

This book also has things to say about place and space in the history of science and medicine. Most obviously, it shows that it is possible, and productive, to take field science seriously as a practice in its own right, not merely as a contrast to, or adaptation of, laboratory research. Fieldwork emerges here as an extraordinarily heterogeneous and complex practice: philosophically, there is a great distance between the relatively controlled, artificial space of Silver Hut and a gasping physiologist fumbling with an alveolar air sampler 3,000 m higher, in the "death zone" of Everest. Practices that would appear to be "natural history" according to the dominant categorizations of "ways of knowing"[64]—such as the gathering of thousands of records of clothing and temperature on an Antarctic expedition—turn out to be useless (and unused) without high-powered computing expertise back in a processing room in Bristol. Meanwhile, a hut in the Antarctic could be even more exclusive, and tents on the side of Everest more segregated, than any elite university laboratory; far from being dangerously porous, the sites of extreme physiology work could exclude the "wrong chaps" (and with them the "wrong" research questions, subjects, and priorities) at least as effectively as traditional laboratory science could.

The exclusivity of extreme physiology research spaces was one factor that maintained the authority of those who worked in them. While the discipline had its disputes—many of them, and some of them fervent and lifelong—there is no evidence that the laboratory, or the iconic biosciences of the twentieth century, had a better claim to truth than the less fashionable discipline of field physiology. Even though extreme physiology experiments may be effectively unreplicable (e.g., if part of the methodology is to climb Everest), this never seems to have been a serious hurdle for publication or authority claims by its participants. Indeed, the repeated failure of reductive mathematical and climate chamber experimental models demonstrated the superiority of the field and the expedition as a site for the production of facts about the natural world and the human body's responses to changing environments. No piece of equipment was taken as proven until it had been used on the mountain or at the poles.[65] The relief expressed by physiologists when the first barometric pressure reading on Everest proved that physicists and their mathematical models were wrong while physiologists and their field studies were right belies nearly a century of prioritization of real-world expertise over models of all kinds.

That this expertise was unevenly distributed is the key focus of chapter 4, which picks apart the ways in which indigenous environmental knowledge and survival technologies could be reinvented as Western or used to demonstrate the superiority of Western explorers and scientists. That chapter and chapter 3 also show the importance of experience and scientific lineage in the formation of expertise. Not only is there a circular process in place whereby those who get opportunities in this unique research field are then much more likely to get further opportunities, but there is also a strong trend in the discipline to appeal to other people's experience through practices of pilgrimage and claims to "genealogies" of science or exploration to demonstrate one's own place in a longer story of adventure and discovery. This trend is layered over the practice—perhaps as old as the Enlightenment, according to Dorinda Outram's account—of using a traveler's body (in this case an explorer's body) as the "locus of authority" when it comes to knowledge about the world; suffering, physical and mental bravery, stoicism and heroism all add authenticity to an "objective" scientific expedition.[66] (And there are also mechanisms to alienate women's bodies—as Naomi Oreskes has so convincingly shown, tasks that are "mere drudgery" for a woman can be interpreted as "heroic stoicism" for a man.)[67] That so much of the methodology of extreme physiology is tacit knowledge, passed on personally or through material objects, is also a form of gatekeeping that maintains the identity and authority of a clique of practitioners.

Material culture is another central focus for this book: objects, whether a tin of pemmican or a gas mask, were used as facts, were traded in exchange for authority or prestige, were passed on as a way of establishing relationships, and were reinvented to disguise their origins. Material objects were often also the major positive outcome of extreme physiology research. The theoretical findings of researchers in extreme environments did have impacts beyond (earthly) exploration: claims, discussed in chapter 5, that the human newborn is born adapted to altitude are part of the reason why leading researchers into respiration science, such as Joseph Barcroft, shifted their attention to the fetal economy; the findings about gasping mountaineers led indirectly to an understanding of the oxygen needs of the newborn, particularly the premature baby, and to designs for new oxygen-providing intensive care incubators (although these devices were radically redesigned when it was discovered that excess oxygen could cause a form of blindness). Likewise, Pugh was not the only extreme physiologist who adapted his research to benefit the sporting community, from the amateur hill walker caught in an unseasonable downpour to the Olympic runner concerned about a track at mid-altitude. The extraordinary fact that a human walked on the moon before a human managed to take a barometric pressure reading on Everest is in part due to the use of mountain-based extreme physiology research by NASA and other organizations.

In other areas, however, extreme physiology failed to draw clear and unequivocal conclusions that could be applied more broadly. Respiration physiology was significantly disrupted by the rather late discovery that Sherpa adaptations are not the same as Andean adaptations, and the hunts for certainty about short-term acclimatization to cold and heat were constant sources of disappointment. This was in part because, as chapter 2 shows, reductive laboratory models were never the priority form of the science and were almost immediately the subject of critique from a body of practitioners who recognized—in some cases, celebrated—a more holistic understanding of the homeostatic regulation of human bodies. In some cases, as Clements has argued for the AMEE, scientists of all stripes struggled to demonstrate the universality of their research findings when their field sites were somewhere as unusual as Everest.[68] But more commonly, physiologists questioned (albeit obliquely) the very principle of simplistic universal theories when it came to the complicated machinery of the human body. Given that the caloric demands of man-hauling versus dog-sledding explorers could differ by 300%, that the wetting of clothing could reduce its insulating properties by 85%, that the peculiarities of individual taste and the psychology of stress could cause climbers with adequate rations to literally starve themselves during a climb, the idea of finding one-size-fits-all models of acclimatization to high, cold, or hot spaces seemed laughable; here were situations in which only the human body, and not a model or a simulation, could "make truth."[69] In part because of the extremity of the stakes, the unusual spaces of the Far South, Far North, and high mountains proved to be valuable research sites for understanding the effect of human individuality on research practices and findings.

These difficulties did not stop extreme physiologists from *trying* to draw universal conclusions: dozens, possibly even hundreds, of studies on cold adaptation were conducted through the twentieth century, even though almost every study started with a literature review suggesting that existing results were conflicting and concluded with a discussion that emphasized the difficulty of drawing conclusions when there were so many confounding factors. One can read the frustration of the researchers desperately seeking a definitive way of answering their deceptively simple question: Do humans adapt to cold environments? (Allan Rogers's experiment on the TAE is a classic example: having settled on what was an apparently simple and robust methodology, he had to wait thirteen years to process the data.) Fatigue, stress, diet, psychological factors, undiagnosed illness, and the use of technology disrupted the research programs of physiologists. Further, physiologists repeatedly faced the challenge of getting subjects in the Arctic and

Antarctic to actually experience cold stress, given the good-quality build-ings and (indigenous-inspired) clothing available to them and their bodies' metabolic responses to hard work. Indeed, several researchers began to sug-gest that these so-called extreme environments were not producing stressed human subjects at all.[70]

But these failures run alongside a series of considerable successes in cre-ating technologies of survival. This book's emphasis on material culture, particularly quotidian technologies such as shoes or sleds or rations, is in part because these technologies were often the only (immediate) certain and tangible outcomes of extreme physiology research. Scientists who would express only tentative and cautious views about even the *possibility* of human adaptation to cold would, at least in private, express extraordinarily confident—not to say dogmatic—views about what style of shoes should be used on an expedition. While this technological innovation has not always been a straightforward story of progress—the analysis of George Mallory's clothing suggested that much older clothing technologies may in some cases be superior to twenty-first-century fabrics—it is demonstrably the case that extreme physiologists saved lives when they applied their theories, expertise, and experimental findings to improving and redesigning material objects. The successful summiting of Everest in 1953 was due in no small measure to the dramatic improvement of oxygen technology since the expeditions of the 1920s and '30s; better rations, in terms of nutritional balance, palat-ability, and overall energy content, have ensured that few Antarctic teams faced the same fate as Scott's polar team in 1911; and if clo measurements and copper men turned out to be unreliable ways to measure thermal com-fort, the studies that surrounded them discovered many useful things about the best clothing for cold and windy climates. At a more fundamental level, thousands of people a year are better enabled to enjoy the outdoors safely by the invention of down clothing, waterproof hiking boots, and gourmet dehydrated camping food. This book's plea to reconsider and broaden our definition of bioprospecting so that it can include these forms of technol-ogy as well as cultural practices of survival is in part an effort to reclaim the indigenous stories of this technology, but also to emphasize the real-world importance, and pervasiveness, of such quotidian science and the technolo-gies it produces.

Clearly, then, the extreme environments of the earth, and the biomedi-cal science done there, have given us both a unique form of scientific prac-tice and an extraordinarily useful and rich resource for understanding twentieth-century science more broadly—as demonstrated by the many themes the preceding chapters have explored. This book has been a story about movement and exchange: bodies (living and dead), samples, ideas, and technology have circulated about the globe. It has also been a story about interdisciplinarity: just as forensics and meteorology meet in a paper about corpses and glaciers, the historical actors discussed in the preceding pages move between disciplines and between spaces of work, blurring boundaries—especially the conceptual one between laboratories and field sites. While these stories may have added complexity to our understanding of twentieth-century biomedical science, the book also provides simplicity: the same names, the same funding bodies, the same physical locations, the same overarching themes and theories crop up time and again across the relatively constrained and sometimes alarmingly inbred world of extreme physiology research. Its stories have also begun to reclaim overshadowed participants, particularly women, porters, statistical analysts, and guides— all the invisible technicians of modern biomedical fieldwork. In doing so, it makes a specific case for a broader archival awareness—it is in published popular exploration books that many extreme physiologists become rounded characters and the details of their experimental procedures can be found (and, of course, it is in the acknowledgments that women may be tucked away). Finally, the book has made the case for the everyday objects of survival and experiment: gloves, pemmican, tents, gas masks, all contain within them the potential to tell extraordinary and explanatory stories about the way that modern science has been practiced around the world.

ACKNOWLEDGMENTS

.

As the product of a series of fixed-term academic positions, this book has ac-
crued more debts than I can admit in a reasonably sized acknowledgments
section. Initial research was made possible by the Wellcome Trust (grant
no. 088204/Z/09/A) and the Isaac Newton Trust, which funded a two-year
research fellowship at the Department of History and Philosophy of Science, University of Cambridge. Here several people read early drafts, helped
with grants and jobs, or offered other support of vital importance to the
book: Nick Hopwood, Lauren Kassell, Simon Schaffer, James Secord, Nicky
Reeves, and all the members of the postdoctoral writing and reading group,
especially Sadiah Qureshi. Further work and writing was made possible by
a Birmingham Fellowship held at the Social Studies of Medicine Unit at the
University of Birmingham. Particular thanks are due here to Jonathan Reinarz and Rebecca Wynter and to my supportive colleagues elsewhere at the
university, including Matthew Hilton, Corey Ross, Kate Nichols, and Sadiah,
again. Thanks to Chris Moores for letting me know about the Conquer Everest board game (fig. 8)!

Crucial research opportunities were provided by a Visiting Fellowship
from the Sydney Centre for the Foundations of Science at the University
of Sydney, and I offer my heartfelt thanks to Warwick Anderson, and to
Sebastián Gil-Riaño, Hans Pols, Sarah Walsh, and Jamie Dunk, as well as to
my co–visiting fellow, Janet Golden. Without this opportunity I would not
have found the words or the motivation to write this book.

I also want to express my thanks, as is always the case, to the archivists
and librarians who made this work possible at the Mandeville Special Collections of the University of California, San Diego, the Royal Geographical Society Archives, the Scott Polar Research Institute, and the University
of Adelaide Rare Books & Special Collections. Thanks for assistance with a

last-minute image search should also go to Bailey at Antarctica New Zealand (fig. 3). Special thanks, too, to Harriet Tuckey and Ingereth Macfarlane, who, as relatives of two of the physiologists discussed here, have given advice, information, and encouragement. No history of extreme physiology could pass without thanks to Professor John West, not only for support but also for the creation of two marvelous historical resources: his book and the archival materials at San Diego.

Other debts are owed to Kate Wood, Christine Baker, and Daisie Huang for a New York writing retreat that broke the back of the first draft. Parts of chapter 2 are drawn, with permission, from my article "Experimental Physiology, Everest and Oxygen: From the Ghastly Kitchens to the Gasping Lung," *British Journal of the History of Science* 46 (2013): 123–47; I thank the anonymous reviewers and editors of that journal, and also those of *Isis*, particularly Bernard Lightman, and of *Social Studies in Science*. Audience members for seminar and conference papers are too numerous to mention: if you have asked me a pertinent question in the past decade, you have probably helped make this book.

Three men saw the start, but tragically not the finish, of this work: two department heads at Cambridge—Peter Lipton and John Forrester—and a colleague from Manchester, John Pickstone.

Finally, as always, to Ben, not just for nearly two decades of company, but also for encouraging me, despite my natural laziness, to explore mountains and glaciers myself.

NOTES

CHAPTER ONE

1. G. E. Fogg, *A History of Antarctic Science* (Cambridge: Cambridge University Press, 1992); John B. West, *High Life: A History of High-Altitude Physiology and Medicine* (New York: Published for the American Physiological Society by Oxford University Press, 1998). For Antarctic medicine, see the works of Henry Raymond Guly, "Medical Aspects of the Expeditions of the Heroic Age of Antarctic Exploration (1895–1922)" (Ph.D. diss., University of Exeter, 2015); Guly, "Human Biology Investigations during the Heroic Age of Antarctic Exploration (1897–1922)," *Polar Record* 50 (2014): 183–91; Guly, "Surgery and Anaesthesia during the Heroic Age of Antarctic Exploration (1895–1922)," *British Medical Journal* 347 (December 17, 2013): f7242; Guly, "Bacteriology during the Expeditions of the Heroic Age of Antarctic Exploration," *Polar Record* 49 (2013): 321–27; Guly, "Medical Comforts during the Heroic Age of Antarctic Exploration," *Polar Record* 49 (2013): 110–17; Guly, "Snow Blindness and Other Eye Problems during the Heroic Age of Antarctic Exploration," *Wilderness & Environmental Medicine* 23 (2012): 77–82; Guly, "Frostbite and Other Cold Injuries in the Heroic Age of Antarctic Exploration," *Wilderness & Environmental Medicine* 23 (2012): 365–70.

2. Simon Schaffer, "The Information Order of Isaac Newton's *Principia Mathematica*" (Hans Rausing Lecture, Uppsala University, Sweden, 2008); Londa Schiebinger, *Plants and Empire: Colonial Bioprospecting in the Atlantic World* (Cambridge, MA: Harvard University Press, 2004); Harold L. Burstyn, "'Big Science' in Victorian Britain: The *Challenger* Expedition (1872–76) and Its *Report* (1881–95)," in *Understanding the Oceans: A Century of Ocean Exploration*, ed. Margaret Deacong, Tony Rice, and Colin Summerhayes (Boca Raton, FL: CRC Press, 2002), 49–55; Dorinda Outram, "On Being Perseus: New Knowledge, Dislocation, and Enlightenment Exploration," in *Geography and Enlightenment*, ed. David N. Livingstone and Charles W. J. Withers (Chicago: University of Chicago Press, 1997), 281–94; Richard Sorrenson, "The Ship as a Scientific Instrument in the Eighteenth Century," *Osiris* 11 (1996): 221–36.

3. Jim Endersby, *Imperial Nature: Joseph Hooker and the Practices of Victorian Science* (Chicago: University of Chicago Press, 2008); Hanna Hodacs, "In the Field: Exploring Nature with Carolus Linnaeus," *Endeavour* 34 (2010): 45–49; Hodacs, "Linnaeans Outdoors: The Transformative Role of Studying Nature 'On the Move' and Outside," *British Journal for the History of Science* 44 (2011): 183–209.

4. Felix Driver, *Geography Militant: Cultures of Exploration and Empire* (Oxford: Oxford University Press, 2001). This interest in exploration is in no way limited to the history of science or medicine. For a good overview of work on just one nation, see Dane Kennedy, "British Exploration in the Nineteenth Century: A Historiographical Survey," *History Compass* 5 (2007): 1879–1900.

5. Robert Peary's claim is still disputed (as was Frederick Cook's of 1908), although 1909 is widely taken as the date of "conquest" of the North Pole; the first absolutely certain visitor was Wally Herbert in 1969.

6. S. Naylor and J. Ryan, eds., *New Spaces of Exploration: Geographies of Discovery in the Twentieth Century* (London: I. B. Tauris, 2009).

7. Shirley V. Scott, "Ingenious and Innocuous? Article IV of the Antarctic Treaty as Imperialism," *Polar Journal* 1 (2011): 51–62.

8. Special issue, *Social Studies of Science* 33, no. 5 (October 2003); Jacob Hamlin, *Oceanographers and the Cold War: Disciplines of Marine Science* (Seattle: University of Washington Press, 2005); Helen M. Rozwadowski, *Fathoming the Ocean: The Discovery and Exploration of the Deep Sea* (Cambridge, MA: Harvard University Press, 2005); Keith R. Benson and Helen M. Rozwadowski, eds., *Extremes: Oceanography's Adventures at the Poles* (Sagamore Beach, MA: Science History Publications/USA, 2007); Gary Kroll, *America's Ocean Wilderness: A Cultural History of Twentieth-Century Exploration* (Lawrence: University of Kansas Press, 2008); Simone Turchetti et al., "On Thick Ice: Scientific Internationalism and Antarctic Affairs, 1957–1980," *History and Technology* 24 (2008): 351–76; Jeremy Vetter, ed., *Knowing Global Environments: New Historical Perspectives on the Field Sciences* (New Brunswick, NJ: Rutgers University Press, 2011).

9. Fogg, *History of Antarctic Science.*

10. Philip W. Clements, *Science in an Extreme Environment: The 1963 American Mount Everest Expedition* (Pittsburgh: University of Pittsburgh Press, 2018).

11. Elena Aronova, Karen S. Baker, and Naomi Oreskes, "Big Science and Big Data in Biology: From the International Geophysical Year through the International Biological Program to the Long Term Ecological Research (LTER) Network, 1957–Present," *Historical Studies in the Natural Sciences* 4 (2010): 183–224.

12. Joanna Radin, *Life on Ice: A History of New Uses for Cold Blood* (Chicago: University of Chicago Press, 2017).

13. Andrew Cunningham and Perry Williams, *The Laboratory Revolution in Medicine* (Cambridge: Cambridge University Press, 2002); Graeme Gooday, "Placing or Replacing the Laboratory in the History of Science?" *Isis* 99 (2008): 783–95; M. Guggenheim, "Laboratizing and De-Laboratizing the World: Changing Sociological Concepts for Places of Knowledge Production," *History of the Human Sciences* 25 (2012): 99–118; Robert E. Kohler, "Lab History: Reflections," *Isis* 99 (2008): 761–68; A. Ophir, "The Place of Knowledge: A Methodological Survey," *Science in Context* 4 (1991): 3–21.

14. Gooday, "Placing or Replacing."

15. Robert E. Kohler, "Practice and Place in Twentieth-Century Field Biology: A Comment," *Journal of the History of Biology* 45 (2012): 579–86; Kohler, "Labscapes: Naturalizing the Lab," *History of Science* 40 (2008): 473–501; Kohler, *Landscapes and Labscapes: Exploring the Lab-Field Border in Biology* (Chicago: University of Chicago Press, 2002); Kohler, "Place and Practice in Field Biology," *History of Science* 40 (2002): 189–210; Henrika Kuklick, "Personal Equations: Reflections on the History of Fieldwork, with Special Reference to Sociocultural Anthropology," *Isis* 102 (2011): 1–33.

16. Stephane Le Gars and David Aubin, "The Elusive Placelessness of the Mont-Blanc Observatory (1893–1909): The Social Underpinnings of High-Altitude Observation,"

Science in Context 22 (2009): 509–31; Thomas F. Gieryn, "City as Truth-Spot," *Social Studies of Science* 36 (2006): 5–38.

17. Raf De Bont, *Stations in the Field: A History of Place-Based Animal Research, 1870–1930* (Chicago: University of Chicago Press, 2015).

18. Antony Adler, "The Ship as Laboratory: Making Space for Field Science at Sea," *Journal of the History of Biology* 47 (2013): 333–62; Sorrenson, "Ship as Instrument."

19. Vanessa Heggie, "Why Isn't Exploration a Science?" *Isis* 105 (2014): 318–34.

20. Bruno J. Strasser, "Collecting, Comparing, and Computing Sequences: The Making of Margaret O. Dayhoff's Atlas of Protein Sequence and Structure, 1954–1965," *Journal of the History of Biology* 43 (2009): 623–60; Strasser, "Laboratories, Museums, and the Comparative Perspective: Alan A. Boyden's Serological Taxonomy, 1925–1962," *Historical Studies in the Natural Sciences* 40 (2010): 533–64; Strasser, "The Experimenter's Museum: GenBank, Natural History, and the Moral Economies of Biomedicine," *Isis* 102 (2011): 60–96.

21. Schiebinger, *Plants and Empire*.

22. Abena Dove Osseo-Asare, "Bioprospecting and Resistance: Transforming Poisoned Arrows into Strophantin Pills in Colonial Gold Coast, 1885–1922," *Social History of Medicine* 21 (2008): 269–90; Osseo-Asare, *Bitter Roots: The Search for Healing Plants in Africa* (Chicago: University of Chicago Press, 2014).

23. Hanne E. F. Nielsen, "Hoofprints in Antarctica: Byrd, Media, and the Golden Guernseys," *Polar Journal* 6 (July 2, 2016): 342–57.

24. Peder Roberts and Dolly Jørgensen, "Animals as Instruments of Norwegian Imperial Authority in the Interwar Arctic," *Journal for the History of Environment and Society* 1 (2016): 65–87.

25. Among the literature on medical geography, of particular relevance to exploration is Frank A. Barrett, "'Scurvy' Lind's Medical Geography," *Social Science & Medicine* 33 (1991): 347–53.

26. Santiago Aragón, "Le rayonnement international de la Société zoologique d'acclimatation: Participation de l'Espagne entre 1854 et 1861," *Revue d'histoire des sciences* 58 (2005): 169–206.

27. Lisbet Koerner, "Purposes of Linnean Travel: A Preliminary Research Report," in *Visions of Empire: Voyages, Botany and the Representation of Nature*, ed. David Miller and Peter Reill (Cambridge: Cambridge University Press, 1999), 117–52. And for the later societies attempting similar things, see Christopher Lever, *They Dined on Eland: The Story of the Acclimatisation Societies* (London: Quiller Press, 1999); K. Anderson, "Science and the Savage: The Linnean Society of New South Wales, 1874–1900," *Cultural Geographies* 5 (1998): 125–43; and Michael Osborne, "Acclimatizing the World: A History of the Paradigmatic Colonial Science," *Osiris* 15 (2000): 135–51.

28. David N. Livingstone, "The Moral Discourse of Climate: Historical Considerations on Race, Place and Virtue," *Journal of Historical Geography* 17 (1991): 413–34.

29. Hans Pols, "Notes from Batavia, the Europeans' Graveyard: The Nineteenth-Century Debate on Acclimatization in the Dutch East Indies," *Journal of the History of Medicine and Allied Sciences* 67 (2012): 120–48.

30. Warwick Anderson, "Immunities of Empire: Race, Disease, and the New Tropical Medicine, 1900–1920," *Bulletin of the History of Medicine* 70 (1996): 94–118; Anderson, "Climates of Opinion: Acclimatization in Nineteenth-Century France and England," *Victorian Studies* 35 (1992): 135–57; David Arnold, ed., *Warm Climates and Western Medicine: The Emergence of Tropical Medicine, 1500–1900* (Amsterdam: Rodopi, 1996); Mark Harrison, *Climates & Constitutions: Health, Race, Environment*

and British Imperialism in India, 1600–1850 (Oxford: Oxford University Press, 1999); Harrison, "'The Tender Frame of Man': Disease, Climate and Racial Difference in India and the West Indies, 1760–1860," *Bulletin of the History of Medicine* 70 (1996): 68–93; Richard Eves, "Unsettling Settler Colonialism: Debates over Climate and Colonization in New Guinea, 1875–1914," *Ethnic and Racial Studies* 28 (2005): 304–30; Michael Joseph, "Military Officers, Tropical Medicine, and Racial Thought in the Formation of the West India Regiments, 1793–1802," *Journal of the History of Medicine and Allied Sciences* 72 (2016): 142–65; Michael Worboys, "The Emergence of Tropical Medicine: A Study in the Establishment of a Scientific Speciality," in *Perspectives on the Emergence of Scientific Disciplines*, ed. Gerard Lemaine (The Hague: De Gruyter Mouton, 1976), 76–98.

31. Livingstone, "Moral Discourse."
32. The picture presented here is Anglocentric; in other European nations, the formation of tropical medicine took slightly different paths, especially when shaped by military rather than civilian medical structures—see Michael Osborne, *The Emergence of Tropical Medicine in France* (Chicago: University of Chicago Press, 2014).
33. Philip D. Curtin, *Death by Migration: Europe's Encounter with the Tropical World in the Nineteenth Century* (Cambridge: Cambridge University Press, 1989).
34. "The school [of tropical medicine] strikes, and strikes effectively, at the root of the principal difficulty of most colonies—disease. It will cheapen government and make it more efficient. It will encourage and cheapen commercial enterprise. It will conciliate and foster the native." Manson, quoted in Michael Worboys, "Tropical Medicine," in *Companion Encyclopaedia of the History of Medicine*, ed. Roy Porter and W. F. Bynum (London: Taylor & Francis, 1993), 512–36.
35. Warwick Anderson, "Geography, Race and Nation: Remapping 'Tropical' Australia, 1890–1930," *Medical History Supplement* 44, S20 (2000): 146–59.
36. The other peak is in the late twentieth century, allied with the rise of "new environmentalism." David N. Livingstone, "Changing Climate, Human Evolution, and the Revival of Environmental Determinism," *Bulletin of the History of Medicine* 86 (2012): 564–95.
37. Warwick Anderson, *The Cultivation of Whiteness: Science, Health, and Racial Destiny in Australia* (Durham, NC: Duke University Press, 2006); Anderson, "Where Every Prospect Pleases and Only Man Is Vile: Laboratory Medicine as Colonial Discourse," *Critical Inquiry* 18 (1992): 506–29.
38. West, *High Life*. John B. West was responsible for setting up the High Altitude Medicine and Physiology collection within the Mandeville Special Collections at the library of the University of California, San Diego ("Mandeville" in these notes).
39. James A. Horscroft et al., "Metabolic Basis to Sherpa Altitude Adaptation," *Proceedings of the National Academy of Sciences* 114 (2017): 6382–87.
40. Juanma Sánchez Arteaga, "Biological Discourses on Human Races and Scientific Racism in Brazil (1832–1911)," *Journal of the History of Biology* 50 (May 2017): 267–314; Marcos Cueto, "Laboratory Styles in Argentine Physiology," *Isis* 85 (1994): 228–46; Cueto, "Andean Biology in Peru: Scientific Styles on the Periphery," *Isis* 80 (1989): 640–58; Jorge Lossio, "Life at High Altitudes: Medical Historical Debates (Andean Region, 1890–1960)" (Ph.D. diss., University of Manchester, 2006); Stefan Pohl-Valero, "¿Agresiones de la altura y degeneración fisiológica? La biografía del 'clima' como objeto de investigación científica en Colombia durante el siglo XIX e inicios del XX," in "Historias alternativas de la fisiología en América Latina," número especial, *Revista Ciencias de la Salud* 13 (2015): 65–83; Pohl-Valero, "'La raza entra por

la boca': Energy, Diet, and Eugenics in Colombia, 1890–1940," *Hispanic American Historical Review* 94 (2014): 455–86.

41. Morgan Seag, "Women Need Not Apply: Gendered Institutional Change in Antarctica and Outer Space," *Polar Journal* 7 (2017): 319–35; Seag, "Equal Opportunities on Ice: Examining Gender and Institutional Change at the British Antarctic Survey, 1975–1996" (master's thesis, University of Cambridge, 2015).

42. University of California, San Diego, Mandeville Special Collections [hereafter Mandeville], Pugh Papers (MSS491), box 42, folder 8, Photo cliché nos. 1925 & 1928; on the disputes, see also James Milledge Papers (MSS455), box 1, folder 20, Diary.

43. For a simple briefer on the linguistic conventions, see "Inuit or Eskimo: Which Name to Use?" at the website of the Alaska Native Language Center, University of Alaska Fairbanks, https://www.uaf.edu/anlc/resources/inuit-eskimo/ (accessed May 2017). Note that *Eskimo* was removed from US federal legislation (and replaced by "Alaska Native") only in 2016. Annie Zak, "Obama Signs Measure to Get Rid of the Word 'Eskimo' in Federal Laws," *Alaska Dispatch News*, May 24, 2016, https://www.adn.com/alaska-news/2016/05/23/obama-signs-measure-to-get-rid-of-the-word-eskimo-in-federal-laws/ (accessed May 2016).

44. Iwan Rhys Morus, "Invisible Technicians, Instrument Makers and Artisans," in *A Companion to the History of Science*, ed. B. Lightman (London: Wiley Blackwell, 2016), 97–110; Steven Shapin, "The Invisible Technician," *American Scientist* 77 (1989): 554–63.

45. Sherry B. Ortner, *Life and Death on Mt. Everest: Sherpas and Himalayan Mountaineering* (Princeton, NJ: Princeton University Press, 1999); Ortner, "Thick Resistance: Death and the Cultural Construction of Agency in Himalayan Mountaineering," *Representations* 59 (1997): 135–62.

46. G. Godin and Roy J. Shephard, "Activity Patterns of the Canadian Eskimo," in *Polar Human Biology: The Proceedings of the SCAR/IUPS/IUBS Symposium on Human Biology and Medicine in the Antarctic*, ed. O. G. Edholm and E. K. Eric Gunderson (London: Heinemann Medical, 1973), 193–215.

47. Animals that live in extreme conditions—not just of heat and cold, but also acidity, alkalinity, etc.

48. Vanessa Heggie, "Experimental Physiology, Everest and Oxygen: From the Ghastly Kitchens to the Gasping Lung," *British Journal for the History of Science* 46 (2013): 123–47.

49. Peder Roberts, "Heroes for the Past and Present: A Century of Remembering Amundsen and Scott," in "Beyond the Limits of Latitude: Reappraising the Race to the South Pole," special issue, *Endeavour* 35 (2011): 142–50; Max Jones, "From 'Noble Example' to 'Potty Pioneer': Rethinking Scott of the Antarctic, c. 1945–2011," *Polar Journal* 1 (2011): 191–206; Jones, *The Last Great Quest: Captain Scott's Antarctic Sacrifice* (Oxford: Oxford University Press, 2003).

CHAPTER TWO

1. Paul Bert, *La pression barométrique: Recherches de physiologie expérimentale* (Paris: G. Masson, 1878), 759–63.

2. Bert, *La pression*, 1105.

3. In the pages before the "Everest" summit, Bert had also used the height of "Mexico" and "Mount Blanc" as comparators.

4. Gabriel Auvinet and Monique Briulet, "El Doctor Denis Jourdanet: Su Vida y Su Obra," *Gaceta médica de México* 140 (2004): 426–29.

198 / Notes to Pages 22–28

5. R. H. Kellogg, "'La Pression Barométrique': Paul Bert's Hypoxia Theory and Its Critics," *Respiration Physiology* 34 (1978): 1–28.

6. Philipp Felsch, *Laborlandschaften: Physiologische Alpenreisen im 19. Jahrhundert* (Göttingen: Wallstein, 2007).

7. Andrew Cunningham and Perry Williams, eds., *The Laboratory Revolution in Medicine* (Cambridge: Cambridge University Press, 1992).

8. William Rostène, "Paul Bert: Homme de science, homme politique," *Journal de la Société de Biologie* 200 (2006): 245–50.

9. Marc Dufour, "Sur le mal de montagne," *Bulletin de la Société médicale de la Suisse Romande* 74 (1874): 72–79, 261–64.

10. Philipp Felsch, "Mountains of Sublimity, Mountains of Fatigue: Towards a History of Speechlessness in the Alps," *Science in Context* 22 (2009): 341–64; Richard Gillespie, "Industrial Fatigue and the Discipline of Physiology," in *Physiology in the American Context, 1850–1940*, ed. Gerald L. Geison (Bethesda, MD: American Physiological Society, 1987), 237–62.

11. Anson Rabinbach, *The Human Motor: Energy, Fatigue and the Origins of Modernity* (Berkeley: University of California Press, 1992).

12. Vanessa Heggie, "Introduction: Special Section—Harvard Fatigue Laboratory," *Journal of the History of Biology* 48 (2015): 361–64.

13. Steven M. Horvath and Elizabeth C. Horvath, *The Harvard Fatigue Laboratory: Its History and Contributions*, International Research Monograph Series in Physical Education (Englewood Cliffs, NJ: Prentice-Hall, 1973).

14. Felsch, *Laborlandschaften.*

15. Camillo Di Giulio and John B. West, "Angelo Mosso's Experiments at Very Low Barometric Pressures," *High Altitude Medicine & Biology* 14 (2013): 78–79.

16. John B. West, *High Life: A History of High-Altitude Physiology and Medicine* (Oxford: Oxford University Press, 1998), 81–82.

17. Angelo Mosso, *Life of Man on the High Alps*, trans. E. Lough Kiesow (London: T. Fisher Unwin, 1898), 308–9.

18. Deborah R. Coen, "The Storm Lab: Meteorology in the Austrian Alps," *Science in Context* 22 (2009): 463–86; Stephane Le Gars and David Aubin, "The Elusive Placelessness of the Mont-Blanc Observatory (1893–1909): The Social Underpinnings of High-Altitude Observation," *Science in Context* 22 (2009): 509–31.

19. David Aubin, "The Hotel That Became an Observatory: Mount Faulhorn as Singularity, Microcosm, and Macro-Tool," *Science in Context* 22 (2009): 365–86.

20. Mosso, *Life of Man*, 267.

21. Felsch, *Laborlandschaften*, 59.

22. Mosso, *Life of Man*, 65.

23. "To my astonishment not one of them spoke of any benefit obtained from inhalations of oxygen. That evening as the guides sat drinking, one of them broke out with the remark that the wine was better than oxygen, and this was repeated by all as a good joke." Mosso, *Life of Man*, 177.

24. Mosso, *Life of Man*, 179.

25. Kellogg, "La Pression Barométrique," 21.

26. See the argument for a circulatory, muscle pressure, and fatigue explanation in Clinton T. Dent, "Can Mount Everest Be Ascended?" *Nineteenth Century* 32 (1892): 604–13.

27. T. G. Longstaff, *Mountain Sickness and Its Probable Causes* (London: Spottiswoode, 1906); Royal Geographical Society Archives [hereafter RGS], EE/98, Minute Books

of the Himalayan Committee, and EE/96, Everest Committee Minutes. See also the description of the conflict between George Ingle Finch, a mountaineer who favored oxygen, and Longstaff in George W. Rodway, "Historical Vignette: George Ingle Finch and the Mount Everest Expedition of 1922: Breaching the 8000-m Barrier," *High Altitude Medicine & Biology* 8 (2007): 68–76.

28. Gordon Douglas et al., "Physiological Observations Made on Pike's Peak, Colorado, with Special Reference to Adaptation to Low Barometric Pressures," *Philosophical Transactions of the Royal Society of London, Series B* 203 (1913): 185–318, Addendum 310.

29. N. Zuntz et al., *Höhenklima und Bergwanderungen in ihrer Wirkung auf den Menschen* (Berlin: Deutsches Verlagshaus, 1906).

30. Hans-Christian Gunga, *Nathan Zuntz: His Life and Work in the Fields of High Altitude Physiology and Aviation Medicine* (New York: Springer Verlag, 2008).

31. West, *High Life*, 92.

32. West, *High Life*, 93; Joseph Barcroft, "The Effect of Altitude on the Dissociation Curve of Blood," *Journal of Physiology* 42 (1911): 44–63.

33. M. P. FitzGerald, "The Changes in the Breathing and the Blood at Various High Altitudes," *Proceedings of the Royal Society of London, Series B* 88 (1913): 351–71.

34. Douglas et al., "Physiological Observations," 186.

35. This is exactly the same argument—"normality"—that was mobilized by the leader of the first extended barometric experimental trials in 1944 (described below). C. S. Houston and R. L. Riley, "Respiratory and Circulatory Changes during Acclimatisation to High Altitude," *American Journal of Physiology* 149 (1947): 565–88. See also the repeated use of the phrase "natural laboratory" to describe the polar regions in O. G. Edholm and E. K. Eric Gunderson, eds., *Polar Human Biology: The Proceedings of the SCAR/IUPS/IUBS Symposium on Human Biology and Medicine in the Antarctic* (London: Heinemann Medical, 1973).

36. Douglas et al., "Physiological Observations," 308.

37. John Hunt, *The Ascent of Everest* (London: Hodder & Stoughton, 1953), 276.

38. Vanessa Heggie, "Experimental Physiology, Everest and Oxygen: From the Ghastly Kitchens to the Gasping Lung," *British Journal for the History of Science* 46 (2013): 123–47.

39. L. G. C. E. Pugh, "The Effects of Oxygen on Acclimatized Men at High Altitude," *Proceedings of the Royal Society of London, Series B* 143 (1954): 17.

40. West, *High Life*, 169.

41. "Obituary: Alexander Mitchell Kellas, D.Sc. (Lond.), Ph.D. (Heidelberg)," *Geographical Journal* 58 (July 1921): 73–75; J. S. Haldane, A. M. Kellas, and E. L. Kennaway, "Experiments on Acclimatisation to Reduced Atmospheric Pressure," *Journal of Physiology* 53 (1919): 181–206.

42. Kellas, A. M., "Dr. Kellas' Expedition to Kamet," *Geographical Journal* 57 (1921): 124–30.

43. T. Howard Somervell, *After Everest: The Experiences of a Mountaineer and Medical Missionary*, 2nd ed. (London: Hodder & Stoughton, 1939), 107.

44. Jeremy Windsor, Roger C. McMorrow, and George W. Rodway, "Oxygen on Everest: The Development of Modern Open-Circuit Systems for Mountaineers," *Aviation, Space, and Environmental Medicine* 79 (A2008): 799–804.

45. These problems were not replicated in the laboratory, as the researchers for the 1953 Everest expedition had to relearn when trying to develop a closed-circuit system in the 1950s. See RGS, EE/75, Report by Campbell Secord to the MRC High Altitude Committee, January 8, 1953: "Two of the runs were done at 250 mm. in the IAM chamber, but these were discontinued when it was realised that conditions were less representative

of a climber at 28,000′ than sea-level tests (no increase of moisture, mass-flow one third of normal, and pressure drop one-ninth of its high-altitude value)."

46. Hugh Ruttledge, "The Mount Everest Expedition, 1933," *Geographical Journal* 83 (1934): 2.

47. T. Howard Somervell, "Note on the Composition of Alveolar Air at Extreme Heights," *Journal of Physiology* 60 (September 4, 1925): 282–85; C. B. Warren, "The Medical and Physiological Aspects of the Mount Everest Expeditions," *Geographical Journal* 90 (August 1937): 126–43.

48. J. B. Haldane et al., "Physiological Difficulties in the Ascent of Mount Everest: Discussion," *Geographical Journal* 65 (January 1925): 16. Kellas said essentially the same thing in A. M. Kellas, "A Consideration of the Possibility of Ascending the Loftier Himalaya," *Geographical Journal* 49 (January 1917): 26–46.

49. R. W. G. Hingston, "Physiological Difficulties in the Ascent of Mount Everest," *Geographical Journal* 65 (1925): 4–16.

50. Harriet Pugh Tuckey, *Everest—The First Ascent: The Untold Story of Griffith Pugh, the Man Who Made It Possible* (London: Rider, 2013), 35–36.

51. See, for example, E. Simons and O. Oelz, "Mont Blanc with Oxygen: The First Rotters," *High Altitude Medicine & Biology* 2 (2001): 545–49.

52. Walt Unsworth, *Everest: The Mountaineering History*, 3rd ed. (London: Bâton Wicks, 2000), 78.

53. Georges Dreyer is sometimes mistakenly rendered as "George," e.g., in Rodway, "Historical Vignette."

54. M. P. FitzGerald and G. Dreyer, *The Unreliability of the Neutral Red Method, as Generally Employed, for the Differentiation of* B. typhosus *and* B. coli, reprinted from *Contributions from the University Laboratory for Medical Bacteriology [Copenhagen] to Celebrate the Inauguration of the State Serum Institute* (Copenhagen: O. C. Olsen & Co., 1902).

55. University of California, San Diego, Mandeville Special Collections [hereafter Mandeville], West Papers (MSS444), box 20, folder 12, letter, West to Dr. Scott Russell, February 1, 1988.

56. E. Hohwu Christensen, "Respiratory Control in Acute and Prolonged Hypoxia," *Proceedings of the Royal Society of London, Series B* 143 (1954): 8–12.

57. The same point was made by Haldane in 1927: J. S. Haldane, "Acclimatisation to High Altitudes," *Physiological Reviews* VII 3 (1927): 363–84.

58. Joseph Barcroft et al., "Observations upon the Effect of High Altitude on the Physiological Processes of the Human Body, Carried out in the Peruvian Andes, Chiefly at Cerro de Pasco," *Philosophical Transactions of the Royal Society of London, Series B* 211 (1923): 351–480; Joseph Barcroft, "Recent Expedition to the Andes for the Study of the Physiology of High Altitudes (BAAS Section of Physiology)," *Lancet* 200 (1922): 685–86.

59. J. S. Milledge, "The Great Oxygen Secretion Controversy," *Lancet* 326 (1985): 1408–11.

60. Milledge, "Great Oxygen Secretion Controversy," 1410.

61. Marcos Cueto, "Andean Biology in Peru: Scientific Styles on the Periphery," *Isis* 80 (1989): 644–46.

62. Joseph Barcroft, *The Respiratory Function of the Blood, Part I: Lessons from High Altitudes* (Cambridge: Cambridge University Press, 1925), 176.

63. Cueto, "Andean Biology."

64. Ancel Keys, "The Physiology of Life at High Altitudes," *Scientific Monthly* 43 (1936): 289.

65. Keys, "Physiology of Life," 281.
66. Sarah Tracy, "The Physiology of Extremes: Ancel Keys and the International High Altitude Expedition of 1935," *Bulletin of the History of Medicine* 86 (2012): 627–60.
67. H. T. Edwards, "Lactic Acid in Rest and Work at High Altitude," *American Journal of Physiology* 116 (1936): note 1, p. 367.
68. This was in no sense an unusual model for exploration funding: the British Arctic expeditions of the early twentieth century sometimes took "paying members" on the promise of an opportunity to hunt polar bears. John Wright, "British Polar Expeditions 1919–39," *Polar Record* 26 (1990): 80.
69. L. G. C. E. Pugh, "Haemoglobin Levels in the British Himalayan Expeditions to Cho Oyu in 1952 and Everest in 1953," *Journal of Physiology* 126 (1954): 38–39.
70. John R. Sutton, "A Lifetime of Going Higher: Charles Snead Houston," *Journal of Wilderness Medicine* 3 (1992): 225–31.
71. H. W. Tilman, *The Ascent of Nanda Devi* (Cambridge: Cambridge University Press, 1937).
72. Charles S. Houston, "Operation Everest: A Study of Acclimatization to Anoxia," *US Naval Medical Bulletin* 46 (1946): 1783–92.
73. H. L. Roxburgh, "Oxygen Equipment for Climbing Mount Everest," *Geographical Journal* 109 (1947): 208. On "flying stress" and tests, see Mark Jackson, "Men and Women under Stress: Neuropsychiatric Models of Resilience During and After the Second World War," in *Stress in Post-War Britain, 1945–85*, ed. Mark Jackson (London: Pickering & Chatto, 2015), 111–29.
74. Roxburgh, "Oxygen Equipment," 208.
75. The Bourdillons concentrated on trying to develop a closed-circuit oxygen system; closed-circuit technology is more complicated than open-circuit technology, as it involves the "recycling" of exhaled breath. Although a closed-circuit system was tested on Everest in 1953, it was the open-circuit system that remained the most popular, and reliable, oxygen technology on the mountain, and it is the only easily available option for mountaineers in the twenty-first century. Windsor, McMorrow, and Rodway, "Oxygen on Everest."
76. RGS, EE/75, draft of *The British Attempt on Everest, 1953*.
77. Houston and Riley, "Respiratory and Circulatory Changes."
78. Zuntz et al., *Höhenklima und Bergwanderungen*, 38.
79. FitzGerald, "Changes in the Breathing."
80. L. G. C. E. Pugh, "Resting Ventilation and Alveolar Air on Mount Everest," *Journal of Physiology* 135 (1957): 604.
81. Pugh, "Resting Ventilation," 606.
82. Pugh, "Resting Ventilation."
83. D. B. Dill and D. S. Evans, "Report Barometric Pressure!" *Journal of Applied Physiology* 29 (1970): 914–16.
84. Many more measurements had been made on the South Col and by using weather balloons in the region. John B. West, "Barometric Pressures on Mt. Everest," *Journal of Applied Physiology* 86 (1999): 1062–66.
85. Pugh, "Resting Ventilation."
86. Mandeville, West Papers (MSS444), box 42, folder 10, "Plan for a combined mountaineering and scientific expedition to Everest."
87. L. G. C. E. Pugh, "Muscular Exercise on Mount Everest," *Journal of Physiology* 141 (1958): 233–61.
88. West, *High Life*, 292.

89. Mandeville, Pugh Papers (MSS491), box 42, folder 10, Himalayan Scientific and Mountaineering Expedition 1960/61, Leader Sir Edmund Hillary, Application for a grant for physiological equipment.

90. Tuckey, *Everest.*

91. Mandeville, Pugh Papers (MSS491), box 42, folder 10, letter, Pugh to Elsner, May 25, 1960.

92. Peter Mulgrew, *No Place for Men* (Auckland: Longman Paul, 1981), 39.

93. L. G. C. E. Pugh, "Science in the Himalaya," *Nature* 191 (1961): 429–30.

94. Hillary's yeti team consisted of himself, American zoologist Larry Swan, zoo director Marlin Perkins, another American climber, John Dienhart, *Statesman* journalist Desmond Doil, climbers George Lowe, Peter Mulgrew, and Pat Barcham, and Gill and Nevison.

95. Mulgrew later died in a plane accident in Antarctica, having taken Ed Hillary's place while Hillary was on a speaking tour in the USA. For an account of Hillary's "cerebral vascular incident," see Mandeville, Milledge Papers (MSS0455), box 1, folder 21, Diary, Sunday, May 7, 1961.

96. For more on the clashes between Hillary and Pugh, see Tuckey, *Everest.* For an overview of the experiments, see L. G. C. E. Pugh, "Physiological and Medical Aspects of the Himalayan Scientific and Mountaineering Expedition," *British Medical Journal* (September 8, 1962): 621–27.

97. As a selection: M. B. Gill et al., "Alveolar Gas Composition at 21,000 to 25,700 Ft. (6400–7830 M)," *Journal of Physiology* 163 (1962): 373–77; M. B. Gill et al., "Falling Efficiency at Sorting Cards during Acclimatisation at 19,000 Ft," *Nature* 203 (1964): 436; M. B. Gill and L. G. C. E. Pugh, "Basal Metabolism and Respiration in Men Living at 5,800 m (19,000 Ft)," *Journal of Applied Physiology* 19 (1964): 949–54.

98. J. S. Milledge, "The Silver Hut Expedition, 1960–1961," *High Altitude Medicine & Biology* 11 (2010): 93–101.

99. Milledge, "Silver Hut Expedition."

100. 7,440 m was the height of the bicycle ergometer on Makalu Col; electrocardiograms were also recorded here, while alveolar gas samples were taken at 7,830 m. The majority of the hut's activities took place at 5,800 m.

101. Mandeville, West Papers (MSS444), box 74, folder 15, letter, Hillary to West, February 23, 1976.

102. Mandeville, West Papers (MSS444), box 10, folder 14, letter, Kellogg to West, January 10, 1981.

103. Although the first woman on the summit of Everest was Japanese, she was followed a few days later by the 1975 Chinese expedition, which had a female deputy leader (Phantog, a Tibetan climber), and undertook at least basic scientific activities on the mountain. Unsworth, *Everest,* appendix 4, 598–99 and chap. 15. See also Monica Jackson and Elizabeth Stark, *Tents in the Clouds: The First Women's Himalayan Expedition* (London: Travel Book Club, 1957).

104. See, for example, Nello Pace, L. Bruce Meyer, and Burton E. Vaughan, "Erythrolysis on Return of Altitude Acclimatized Individuals to Sea-Level," *Journal of Applied Physiology* 9 (1956): 141–44; Philip W. Clements, *Science in an Extreme Environment: The 1963 American Mount Everest Expedition* (Pittsburgh: University of Pittsburgh Press, 2018).

105. Unsworth, *Everest,* 461–62.

106. P. Cerretelli, "Limiting Factors to Oxygen Transport on Mount Everest," *Journal of Applied Physiology* 40 (1976): 658–67.

107. R. F. Fletcher, "Birmingham Medical Research Expeditionary Society 1977 Expedition: Signs and Symptoms," *Postgraduate Medical Journal* 55 (1979): 461–63.

108. West, *High Life*, 328–30.

109. John B. West et al., "Pulmonary Gas Exchange on the Summit of Mount Everest," *Journal of Applied Physiology* 55 (1983): 678–87.

110. Mandeville, West Papers (MSS444), box 15, folder 32, letter, West to Sheldon Shultz, February 11, 1982.

111. Mandeville, West Papers (MSS444), box 15, folder 32, letter, West to Sheldon Schultz, February 11, 1982, p. 2.

112. Houston and Riley, "Respiratory and Circulatory Changes," 566. West had refuted this suggestion at length in 1962: John B. West, "Diffusing Capacity of the Lung for Carbon Monoxide at High Altitude," *Journal of Applied Physiology* 17 (1962): 421–26.

113. Mandeville, West Papers (MSS444), box 8, folder 27, letter, West to Charles S. Houston, September 11, 1987.

114. John R. Sutton et al., "Operation Everest II: Oxygen Transport during Exercise at Extreme Simulated Altitude," *Journal of Applied Physiology* 64 (1988): 1309–21.

115. Notably, the volunteer subjects were still all male, and all White. Jean-Paul Richalet, "Operation Everest III: COMEX '97," *High Altitude Medicine & Biology* 11 (2010): 121–32.

116. J. M. van der Kaaij et al., "Research on Mount Everest: Exploring Adaptation to Hypoxia to Benefit the Critically Ill Patient," *Netherlands Journal of Critical Care* 15 (2011): 241.

117. Mandeville, West Papers (MSS444), box 8, folder 4, letter, West to Hackett, December 5, 1988.

118. G. Savourey et al., "Are the Laboratory and Field Conditions Observations of Acute Mountain Sickness Related?" *Aviation, Space, and Environmental Medicine* 68 (1997): 895–99.

119. John T. Reeves et al., "Operation Everest II: Preservation of Cardiac Function at Extreme Altitude," *Journal of Applied Physiology* 63 (1987): 531.

120. See, for example, Mandeville, West Papers (MSS444), box 8, folder 27, letter, West to Charles Houston, September 11, 1987.

121. See, for example, the anecdotes, personal stories, detailed images, and technical specifications in T. D. Bourdillon, "The Use of Oxygen Apparatus by Acclimatized Men," *Proceedings of the Royal Society of London, Series B* 143 (1954): 24–32. But compare this account with the vastly more detailed accounts of the Silver Hut expedition in articles written by Ward for a nonscientific journal: Michael Ward, "Himalayan Scientific Expedition 1960–61 (A Himalayan Winter, Rakpa Peak, Ama Dablam, Makalu)," *Alpine Journal* 66 (1961): 343–64; and Michael Ward, "The Descent from Makalu, 1961, and Some Medical Aspects of High Altitude Climbing," *Alpine Journal* 68 (1963): 11–19.

122. Tuckey, *Everest*, 94. See reports in RGS, EE/90, "Report on Visit to Swiss Foundation for Alpine Research September 22–15, 1952"; and John B. West, "Times Past: Failure on Everest: The Oxygen Equipment of the Spring 1952 Swiss Expedition," *High Altitude Medicine & Biology* 4 (2003): 39–43.

123. J. E. Cotes, "Ventilatory Capacity at Altitude and Its Relation to Mask Design," *Proceedings of the Royal Society of London, Series B* 143 (1954): 32–39.

124. See the letter from Glaxo's advertising department to Hinks, March 23, 1922, in RGS, EE/17/1. Sadly, as Hinks wrote back, "I can . . . tell you privately and not for

publication that I heard from the Chief of the Expedition that most of the members did not like it at all."

125. Mandeville, Hornbein Papers (MSS669), box 31, folder 6, letter, Dyhrenfurth to Mr. Minot Dole, June 9, 1961.

126. "Beyond the satisfaction of contributing to the first American attempt to climb the highest mountain in the world, there is little that can be offered in return for your assistance. It is my hope that we can climb the mountain with oxygen equipment of American design and manufacture not just because it is 'American,' but also because it is superior to anything available elsewhere." Mandeville, Hornbein Papers (MSS669), box 31, folder 7, letter, Hornbein to Harry L. Daulton, January 2, 1961.

127. Mandeville, Hornbein Papers (MSS669), box 31, folder 7, Report by Hornbein on Development of Oxygen Masks, c. 1961.

128. Mandeville, Hornbein Papers (MSS669), box 31, folder 7, letter, Hornbein to Dyhrenfurth, January 12, 1961.

129. Mandeville, Hornbein Papers (MSS669), box 31, folder 6, letter, Hornbein to Dyhrenfurth, July 30, 1961.

130. Dass Deepak and G. Bhaumik, "The Silver Hut Experiment," Science Reporter (November 2012), http://nopr.niscair.res.in/bitstream/123456789/15016/1/SR%2049 %2811%29%2056-57.pdf (accessed June 25, 2014).

131. Bruno Latour, The Pasteurisation of France, trans. Alan Sheridan and John Law (Cambridge, MA: Harvard University Press, 1988).

132. Deepak and Bhaumik, "Silver Hut Experiment."

133. John B. West, "Letter from Chowri Kang," High Altitude Medicine & Biology 2 (2001): 311–13.

134. Similar practices had been undertaken by British Everest teams, using the Alps as a "staging post" for their oxygen equipment, in both the 1930s and 1950s. RGS, EE/54, Report on the Tests with an Oxygen Apparatus in the Alps, c. 1937, and EE/90/12 (various; correspondence with the Swiss).

135. With thanks to the Caudwell Xtreme Everest Expeditions team—particularly Andrew Murray and Mike Grocott—for allowing me to attend a planning meeting.

CHAPTER THREE

1. John B. West, Everest: The Testing Place (New York: McGraw-Hill, 1985), 117.

2. John B. West, "American Medical Research Expedition to Everest, 1981," Himalayan Journal 39 (1981–82): 25.

3. Sherry B. Ortner, Life and Death on Mt. Everest: Sherpas and Himalayan Mountaineering (Princeton, NJ: Princeton University Press, 1999), 61; West, Everest: The Testing Place, 100; Walt Unsworth, Everest: The Mountaineering History, 3rd ed. (London: Bâton Wicks, 2000), 618.

4. The same was true for nonscientist explorers: see the analysis of group formation in John Wright, "British Polar Expeditions 1919–39," Polar Record 26 (1990): 77–84.

5. University of California, San Diego, Mandeville Special Collections [hereafter Mandeville], Hornbein Papers (MSS669), box 33, folder 2, letter, Tom to "Barrel" Bishop, January 8, 1965; Wright, "British Polar Expeditions"; see also Peder Roberts, The European Antarctic: Science and Strategy in Scandinavia and the British Empire (New York: Palgrave Macmillan, 2011), particularly chapter 4.

6. Mandeville, Pugh Papers (MSS491), box 8, folder 29, letter, C. L. Levere to Pugh, August 2, 1958.

7. For multiple examples relating to Antarctic medicine and physiology, see Henry Raymond Guly, "Medical Aspects of the Expeditions of the Heroic Age of Antarctic Exploration (1895–1922)" (Ph.D. diss., University of Exeter, 2015).

8. John B. West, "George I. Finch and His Pioneering Use of Oxygen for Climbing at Extreme Altitudes," *Journal of Applied Physiology* 94 (2003): 1702–13.

9. M. P. FitzGerald and J. S. Haldane, "The Normal Alveolar Carbonic Acid Pressure in Man," *Journal of Physiology* 32 (1905): 486–94.

10. John B. West, *High Life: A History of High-Altitude Physiology and Medicine* (New York: Published for the American Physiological Society by Oxford University Press, 1998), 128.

11. West, *High Life*, 131.

12. Nea Morin, *A Woman's Reach: Mountaineering Memoirs* (London: Eyre & Spottiswoode, 1968), 229–30.

13. Frederic Jackson and Hywel Davies, "The Electrocardiogram of the Mountaineer at High Altitude," *British Heart Journal* 22 (1960): 671–85.

14. E. S. Williams, "Sleep and Wakefulness at High Altitudes," *British Medical Journal* 1 (January 24, 1959): 197.

15. John V. Pickstone, "Museological Science? The Place of the Analytical/Comparative in Nineteenth-Century Science, Technology and Medicine," *History of Science* 31 (1994): 111–38.

16. The organizer of this expedition, Edward S. Williams, wrote several times to Pugh (sending him a research proposal for the expedition) for advice on scientific work at altitude. Mandeville, Pugh Papers (MSS491), box 10, folder 33.

17. West, *Everest: The Testing Place*, 30. It is a sign that things were changing that West is clear that there needs to be an apology for an all-male team and openly notes, "Some will brand our attitude as ultraconservative or even chauvinistic."

18. Polly G. Nicely and Judith K. Childers, "Mt. Everest Reveals Its Secrets to Medicine and Science: A Report on the 1981 American Medical Research Expedition to Everest," *Journal of the Indiana State Medical Association* 75 (1982): 704–8.

19. Nicely and Childers, "Mt. Everest Reveals," 705.

20. M. C. Shelesnyak, "The History of the Arctic Research Laboratory, Point Barrow, Alaska," *Arctic* 1 (1948): 97–106.

21. Mary C. Lobban, "Cambridge Spitsbergen Physiological Expedition, 1953," *Polar Record* 48 (1954): 151–61.

22. Ann M. Savours, "Obituary: Mary C. Lobban," *Polar Record* 21 (1983): 403.

23. Julie Clayton, *MRC National Institute for Medical Research: A Century of Science for Health* (London: MRC, 2014), 253.

24. Helen E. Ross, "Sleep and Wakefulness in the Arctic under an Irregular Regime," in *Biometeorology: Proceedings of the Second International Bioclimatological Conference (1960)*, ed. S. W. Tromp (Oxford: Pergamon Press, 1962), 394.

25. Beau Riffenburgh, ed., *Encyclopaedia of the Antarctic*, vol. 1 (London: Routledge, 2007), 1094.

26. Colin Bull, "Behind the Scenes: Colin Bull Recalls His 10-Year Quest to Send Women Researchers to Antarctica," *Antarctic Sun*, November 13, 2009, https://antarcticsun.usap.gov/features/contentHandler.cfm?id=1955 (accessed June 2017). See also the interview with Colin Bull by Brian Shoemaker, conducted as part of the Polar Oral History Programme (2007), at http://hdl.handle.net/1811/28580 (accessed July 2017).

27. Felicity Aston, "Women of the White Continent," *Geographical* 77 (2005): 26–30.

28. Jennifer Keys and Henry Guly, "The Medical History of South Georgia," *Polar Record* 45 (2009): 270.

29. Aston, "Women of the White Continent."

30. Ove Wilson, "Human Adaptation to Life in Antarctica," in *Biogeography and Ecology in Antarctica*, ed. Van Mieghem and P. van Oye (The Hague: W. Junk, 1965), 732.

31. Morgan Seag, "Women Need Not Apply: Gendered Institutional Change in Antarctica and Outer Space," *Polar Journal* 7 (2017): 319–35.

32. Lowe wrote an autobiographical account of his exploration in the 1950s: *From Everest to the South Pole* (New York: St. Martin's Press, 1961).

33. David Kaiser, *How the Hippies Saved Physics: Science, Counterculture and the Quantum Revival* (London: Norton, 2011).

34. West, *Everest: The Testing Place*, 23.

35. The British Antarctic Survey was still the Falklands Islands Dependencies Survey, and these negotiations were led by its Deputy/Acting Director (1955–58), Sir Raymond Priestley. O. G. Edholm, "Medical Research by the British Antarctic Survey," *Polar Record* 12 (1965): 575–82.

36. Mandeville, West Papers (MSS444), box 14, folder 7, reply to letter, Kurt Papenfus, Aspen Valley Hospital, to West, April 21, 1993.

37. For a review of revisionist histories, see Max Jones, "From 'Noble Example' to 'Potty Pioneer': Rethinking Scott of the Antarctic, c. 1945–2011," *Polar Journal* 1 (2011): 191–206.

38. Max Jones, *The Last Great Quest: Captain Scott's Antarctic Sacrifice* (Oxford: Oxford University Press, 2004).

39. L. G. Halsey and M. A. Stroud, "100 Years since Scott Reached the Pole: A Century of Learning about the Physiological Demands of Antarctica," *Physiological Reviews* 92 (2012): 521–36.

40. M. A. Stroud, "Nutrition and Energy Balance on the 'Footsteps of Scott' Expedition 1984–86," *Human Nutrition: Applied Nutrition* 41 (1987): 426–33.

41. Sally Smith Hughes, "Interview Transcript: Will Siri," Bancroft Library, University of California, Berkeley, 1980, 70, http://digitalassets.lib.berkeley.edu/rohoia/ucb/text /nuclearmedicine00lawrrich.pdf (accessed May 2015).

42. Henry Raymond Guly, "Human Biology Investigations during the Heroic Age of Antarctic Exploration (1897–1922)," *Polar Record* 50 (2014): 183–91.

43. "Nello Pace Biographical Material," White Mountain Research Center, UCLA Institute of the Environment and Sustainability, http://www.wmrc.edu/gifts/pace-bio .html (accessed March 2016).

44. Much of the information in this section has been taken from previously collected oral histories: Anna Berge and Nello Pace, "Human Radiation Studies: Remembering the Early Years—Oral History of Physiologist Nello Pace Ph.D.," United States Department of Energy—Office of Human Radiation Experiments—DOE/EH-0476, June 1995, www.iaea.org/inis/collection/NCLCollectionStore/_Public/27/059/27059281 .pdf (accessed December 2017); Smith Hughes, "Interview Transcript: Will Siri."

45. Mandeville, Pugh Papers (MSS491), box 39, folder 12, letter, Pugh to Professor Brown, April 1955.

46. Mandeville, Pugh Papers (MSS491), box 39, folder 12, letter, E. H. Eckelmeyer to Edholm, April 17, 1957.

47. Siri, William E., and Ann Lage, *William E. Siri: Reflections on the Sierra Club, the Environment and Mountaineering, 1950s–1970s*, Sierra Club History Series (Berkeley: Regional Oral History Office, The Bancroft Library, University of California, 1979), 252.

48. Smith Hughes, "Interview Transcript: Will Siri," 252.

49. Smith Hughes, "Interview Transcript: Will Siri," 71.

50. Mandeville, Pugh Papers (MSS491), box 39, folder 12, letter, Pugh to Pace, July 3, 1957.

51. Mandeville, Pugh Papers (MSS491), box 39, various.

52. "The participation of Dr Pugh on this project is regarded as being in the interest of international cooperation and would undoubtedly result in significant contributions to our knowledge of the effects of polar environments on man." Mandeville, Pugh Papers (MSS491), box 39, folder 12, letter, E. H. Eckelmeyer to Edholm, April 17, 1957.

53. Mandeville, Pugh Papers (MSS491), box 39, folder 14, letter, Pugh to Halve Carlson, April 11, 1957.

54. "Jim Adam" is Major James Adam of the Royal Army Medical Corps. Mandeville, Pugh Papers (MSS491), box 39, folder 14, letter, Pugh to Halve Carlson, April 11, 1957.

55. British readers of a certain age may recognize Dr. Wolff as the judge and later presenter of the competitive engineering game show *The Great Egg Race* (1979–86). Clayton, *MRC National Institute*, 294.

56. Lowe, *From Everest*, 161.

57. Scott Polar Research Institute [hereafter SPRI], Vivian Fuchs Papers (GB 15), ms1536/2, "On Being 'Imped.'"

58. SPRI, Vivian Fuchs Papers (GB 15), ms1536/2, "On Being 'Imped.'"

59. SPRI, Vivian Fuchs Papers (GB 15), ms1536/2, "On Being 'Imped.'"

60. SPRI, Vivian Fuchs Papers (GB 15), ms1536/2, diary, January 28, 1959.

61. Mandeville, Pugh Papers (MSS491), box 3, folder 8; A. F. Rogers and R. J. Sutherland, *Antarctic Climate, Clothing and Acclimatization: Final Scientific Report* (Bristol: Bristol University Department of Physiology, 1971), section IV.7.

62. Mandeville, Pugh Papers (MSS491), box 3, folder 8, diary, January 4, 1958.

63. Rogers and Sutherland, *Antarctic Climate*, section P.II.3.

64. A. B. Blackburn, "Medical Research at Plateau Station," *Antarctic Journal of the United States* 3 (December 1968): 237–39.

65. Mandeville, Pugh Papers (MSS491), box 39, folder 12, International Physiological Expedition to Antarctica.

66. Blackburn, "Medical Research"; R. I. Adam and W. R. Stanmeyer, "Effects of Prolonged Antarctic Isolation on Oral and Intestinal Bacteria," *Oral Surgery, Oral Medicine, Oral Pathology* 13 (1960): 117–20.

67. Mandeville, Pugh Papers (MSS491), box 39, folder 12, transcript of a program "Man below Zero," broadcast Sunday, June 29, 1958, on KNX LA and KCBS San Francisco.

68. Mandeville, Pugh Papers (MSS491), box 3, folder 8, diary, February 19, 1958. Rather tellingly, the entry for February 19 about killing and analyzing seals is followed on February 22 by an entry "Looked up seals in volume of Ency. Brit to find no mention of their blood." This may be what inspired Pugh to set up his own study.

69. T. Howard Somervell, "Note on the Composition of Alveolar Air at Extreme Heights," *Journal of Physiology* 60 (1925): 282–85.

70. Raymond Greene, "Observations on the Composition of Alveolar Air on Everest, 1933," *Journal of Physiology* 82 (1934): 481–85.

71. L. G. C. E. Pugh, "Muscular Exercise on Mount Everest," in *High Altitude Physiology: Benchmark Papers in Human Physiology*, ed. John B. West (Stroudsburg, PA: Hutchinson Ross, 1981), 79–81.

72. West, *Everest: The Testing Place*, 117.

73. West, *Everest: The Testing Place*, 118.

74. D. S. Matthews et al., "Some Effects of High-Altitude Climbing; Investigations Made on Climbers of the British Kangchenjunga Reconnaissance Expedition, 1954," *British Medical Journal* 1 (March 26, 1955): 769.

75. Operation Snuffles was part of the US Operation Deepfreeze IV and included a "British observer" on the icebreaker USS *Staten Island*: Raymond Priestley, himself an Antarctic explorer and scientist and then the director of the British Antarctic Survey. "Biological and Medical Research Based on USS *Staten Island*, Antarctica, 1958–59," *Polar Record* 10 (1960): 146–48.

76. "Biological and Medical Research," 146–47.

77. O. G. Edholm and E. K. Eric Gunderson, eds., *Polar Human Biology: The Proceedings of the SCAR/IUPS/IUBS Symposium on Human Biology and Medicine in the Antarctic* (London: Heinemann Medical, 1973).

78. R. M. Lloyd, "Ketonuria in the Antarctic: A Detailed Study," *British Antarctic Survey Bulletin* 20 (1969): 59–68.

79. D. L. Easty, A. Antonis, and I. Bersohn. "Adipose Fat Composition in Young Men in Antarctica," *British Antarctic Survey Bulletin* 13 (1967): 41–45.

80. Mandeville, Pugh Papers (MSS491), box 1, folder 3, report of a meeting of the scientific staff, December 10, 1957. This MRC meeting was held in the absence of Pugh, who was still in the Antarctic.

81. Rogers and Sutherland, *Antarctic Climate*, methods sections.

82. Rogers and Sutherland, *Antarctic Climate*.

83. Rogers and Sutherland, *Antarctic Climate*, 1.

84. The literature here is huge, but see, for example, A. G. Davis, "Seasonal Changes in Body Weight and Skinfold Thickness," *British Antarctic Survey Bulletin* 19 (1969): 75–81; I. F. G. Hampton, "Local Acclimatisation of the Hands to Prolonged Cold Exposure in the Antarctic," *British Antarctic Survey Bulletin* 19 (1969): 9–56; J. N. Norman, "Cold Exposure and Patterns of Activity at a Polar Station," *British Antarctic Survey Bulletin* 6 (1965): 1–13.

85. Early in the century some explorers had gone so far as to suggest that White bodies might actually have an advantage in the polar regions: the American Arctic explorer Robert Peary claimed that for winter expeditions he preferentially recruited blondes, believing them to be more resistant to the "absence of the actinic or the physiological affects of the sun's rays." Robert E. Peary, *Secrets of Polar Travel* (New York: Century Co., 1917; reprint, Elibrion Classics, 2007), 52.

86. G. M. Brown et al., "The Circulation in Cold Acclimatization," *Circulation* 9 (1954): 813–22.

87. Edholm and Gunderson, *Polar Human Biology*.

88. For a closer look at the internal politics of expeditions, particularly the relationship between Pugh and Hillary, see Harriet Tuckey, *Everest—The First Ascent: The Untold Story of Griffith Pugh, the Man Who Made It Possible* (London: Rider, 2013).

89. There is an extremely comprehensive review of studies in E. K. Eric Gunderson, ed., *Human Adaptability to Antarctic Conditions*, Antarctic Research Series, vol. 22 (Washington, DC: American Geophysical Union, 1974).

90. Vivian Fuchs, foreword to *Man in the Antarctic: The Scientific Work of the International Biomedical Expedition to the Antarctic (IBEA)*, by Jean Rivolier (London: Taylor & Francis, 1988), xvi.

91. Rivolier, *Man in the Antarctic*, 3.

92. Rivolier, *Man in the Antarctic*, 150.

93. Rivolier, *Man in the Antarctic*, 12.

94. Anthony J. W. Taylor and Iain A. McCormick, "Human Experimentation during the International Biomedical Expedition to the Antarctic (IBEA)," *Journal of Human Stress* 11 (1985): 162.

95. Rivolier, *Man in the Antarctic*, 81–82.

96. L. A. Palinkas and D. Browner, "Stress, Coping and Depression in US Antarctic Program Personnel," *Antarctic Journal of the United States* 26 (1991): 240–41.

97. Siri and Lage, *William E. Siri*, 249.

98. Philip W. Clements, *Science in an Extreme Environment: The 1963 American Mount Everest Expedition* (Pittsburgh: University of Pittsburgh Press, 2018); see also the extensive report by James T. Lester, *Behavioral Research during the 1963 American Mount Everest Expedition* (Final Report September 1964), www.dtic.mil/dtic/tr/fulltext/u2/607336.pdf (accessed June 2018).

99. University of Adelaide Rare Books & Special Collections [hereafter Adelaide], W. V. Macfarlane Papers 1947–1985 (MS0006), F2/37, letter, Dr. P. G. Law, March 11, 1980.

100. Mandeville, Pugh Papers (MSS491), box 3, folder 8, diary, January 5, 1958.

101. Lowe, *From Everest*, 155.

102. See, for example, debates over airing "dirty linen" in Mandeville, Hornbein Papers (MSS669), box 32, folder 5, letter, Norman [Dyhrenfurth] to Tom H., November 11, 1963.

103. SPRI, Vivian Fuchs Papers (GB 15), ms1536/2, diary, April 24–25, 1957.

104. SPRI, Vivian Fuchs Papers (GB 15), ms1536/2, diary, April 24–25, 1957.

105. E. F. Adolph, ed., *Physiology of Man in the Desert* (New York: Interscience Publishers, 1947).

106. Lowe, *From Everest*, 35.

107. Wilfred Noyce, *South Col: The Personal Account of One Man's Adventures on Everest* (London: Reprint Society, 1955), 125; John Hunt, *The Ascent of Everest* (London: Hodder & Stoughton, 1953), 115.

108. Lowe, *From Everest*, 36.

109. Noyce, *South Col*, 87.

110. Cf. this quote in relation to one of Chris Bonington's expeditions of the early 1970s: "The frequency of the Face expeditions was such that each camp site was littered with the bric-á-brac of previous expeditions—tent frames and platforms (often damaged), oxygen cylinders, spare rope: all the mountain excreta of our modern consumer technology." Unsworth, *Everest*, 438.

111. Edmund Hillary and Desmond Doig, *High in the Thin Cold Air* (London: Hodder & Stoughton, 1963), 29.

112. Ortner, *Life and Death*.

113. P. J. Capelotti, "Extreme Archaeological Sites and Their Tourism: A Conceptual Model from Historic American Polar Expeditions in Svalbard, Franz Josef Land and Northeast Greenland," *Polar Journal* 2 (2012): 236–55.

114. Noel Barber, *The White Desert* (London: Hodder & Stoughton, 1958), 107.

115. Christy Collis, "Walking in Your Footsteps: 'Footsteps of the Explorers' Expeditions and the Contest for Australian Desert Space," in *New Spaces of Exploration: Geographies of Discovery in the Twentieth Century*, ed. S. Naylor and J. R. Ryan (London: I. B. Tauris & Co., 2009), 222–40.

116. Personal communication, explorer X (anonymized), June 2011.

117. Smith Hughes, "Interview Transcript: Will Siri," 72.

118. L. G. C. E. Pugh, "Carbon Monoxide Content of the Blood and Other Observations on Weddell Seals," *Nature* 183 (1959): 74–76; Pugh, "Carbon Monoxide Hazard in Antarctica," *British Medical Journal* 1 (January 24, 1959): 192–96.

119. Innes M. Keighren, "A Scot of the Antarctic: The Reception and Commemoration of William Speirs Bruce" (master's thesis, University of Edinburgh, 2003), 48–49. See also Beau Riffenburgh, *The Myth of the Explorer: The Press, Sensationalism and Geographical Discovery* (London: Belhaven Press, 1993).

CHAPTER FOUR

1. From Bowers's reckoning of the sledge weights for the first winter journey: Apsley Cherry-Garrard, *The Worst Journey in the World 1910–13*, vol. 1 (London: Constable and Co., 1922), 230–31.

2. Henry Raymond Guly, "Bacteriology during the Expeditions of the Heroic Age of Antarctic Exploration," *Polar Record* 49 (2013): 321–27, table 1.

3. A. B. Blackburn, "Medical Research at Plateau Station," *Antarctic Journal of the United States* III (1968): 237. On the overlap of kit, see Mike Parsons and Mary B. Rose, *Invisible on Everest: Innovation and the Gear Makers* (London: Old City Publishing, 2002).

4. Royal Geographical Society Archives [hereafter RGS], EE/16/1/6, letter, Simpson to Hinks, February 1, 1921.

5. B. Charnley, "Arguing over Adulteration: The Success of the Analytical Sanitary Commission," *Endeavour* 32 (2008): 129–33; S. D. Smith, "Coffee, Microscopy, and the *Lancet*'s Analytical Sanitary Commission," *Social History of Medicine* 14 (2001): 171–97.

6. Arthur J. Ray, "The Northern Great Plains: Pantry of the Northwestern Fur Trade, 1774–1885," *Prairie Forum* 9 (1984): 270–71.

7. For a broader environmental history take on pemmican's role in North America, see George Colpitts, *Pemmican Empire: Food, Trade, and the Last Bison Hunts in the North American Plains, 1780–1882* (Cambridge: Cambridge University Press, 2015).

8. Vanessa Heggie, "Rationalised and Rationed: Food and Health," in *Cultural History of Medicine: Age of Empire, 1800–1920*, ed. J. Reinarz (London: Bloomsbury, forthcoming).

9. For a non-European example, see Gail Borden, *The Meat Biscuit: Invented, Patented, and Manufactured* (New York, 1853).

10. F. Galton, *The Art of Travel*, 1st ed. (London: John Murray, 1855), 49–50.

11. Robert E. Peary, *Secrets of Polar Travel* (New York: Century Co., 1917; reprint, Elibrion Classics, 2007), 83.

12. Vivian Fuchs, "Sledging Rations of the Falkland Islands Dependencies Survey, 1948–50," *Polar Record* 6 (1952): 511.

13. University of California, San Diego, Mandeville Special Collections [hereafter Mandeville], Pugh Papers (MSS491), box 10, folder 9, "The South Georgia Survey 1953–4, VIII sledging rations," n.d.

14. Londa Schiebinger, *Plants and Empire: Colonial Bioprospecting in the Atlantic World* (Cambridge, MA: Harvard University Press, 2004); Abena Dove Osseo-Asare, "Bioprospecting and Resistance: Transforming Poisoned Arrows into Strophantin Pills in Colonial Gold Coast, 1885–1922," *Social History of Medicine* 21 (2008): 269–90.

15. Raf De Bont, "'Primitives' and Protected Areas: International Conservation and the 'Naturalization' of Indigenous People, ca. 1910–1975," *Journal of the History of Ideas* 76 (2015): 215–36.

16. Shane Greene, "Indigenous People Incorporated? Culture as Politics, Culture as Property in Pharmaceutical Bioprospecting," *Current Anthropology* 45 (2004): 211–37; John Merson, "Bio-Prospecting or Bio-Piracy: Intellectual Property Rights and Biodiversity in a Colonial and Postcolonial Context," *Osiris* 15 (2000): 282–96.

17. J. T. Kenny, "Claiming the High Ground: Theories of Imperial Authority and the British Hill Stations in India," *Political Geography* 16 (1997): 655–73.

18. Matthew Farish, "The Lab and the Land: Overcoming the Arctic in Cold War Alaska," *Isis* 104 (2013): 1–29.

19. Michael Ward, "The Height of Mount Everest," *Alpine Journal* 100 (1995): 30–33.

20. Kapil Raj, *Relocating Modern Science. Circulation and the Construction of Knowledge in South Asia and Europe, 1650–1900* (London: Routledge, 2007).

21. Pundits, sometimes rendered as "pandits," were particularly used to gain access to territories that were closed to British explorers or hostile to the idea of a British survey and surveillance. Kapil Raj, "When Human Travellers Become Instruments," in *Instruments, Travel and Science*, ed. Marie Noëlle Bourguet, Christian Licoppe, and H. Otto Sibum (London: Routledge, 2002), 156–88.

22. T. G. Longstaff, "A Mountaineering Expedition to the Himalaya of Garhwal," *Geographical Journal* 31 (1908): 364.

23. "As an instance of the value of local native evidence, I may mention that Mr. J. S. Ward, of the Rifle Brigade, told me that less than three months later our route was pointed out to him as lying over the spurs to the west of Dunagiri, along a shepherd's summer track." Longstaff, "Mountaineering Expedition," 367.

24. Clements R. Markham, *The Lands of Silence* (Cambridge: Cambridge University Press, 1921), 214.

25. Longstaff, "Mountaineering Expedition," 364.

26. H. T. Morshead, "Report on the Expedition to Kamet, 1920," *Geographical Journal* 57 (1921): 219.

27. George W Rodway, "Prelude to Everest: Alexander M. Kellas and the 1920 High Altitude Scientific Expedition to Kamet," *High Altitude Medicine & Biology* 5 (2004): 364–79; Mandeville, West Papers (MSS444), box 70, folder 18, offprint of Paul Geissler, "Alexander M Kellas," *Deutsche Alpenzeitung* 30 (1935): 103–10.

28. A. M. Kellas, "Dr. Kellas' Expedition to Kamet," *Geographical Journal* 57 (1921): 128.

29. Mandeville, West Papers (MSS444), box 70, folder 18, offprint of Paul Geissler, "Alexander M Kellas," *Deutsche Alpenzeitung* 30 (1935): 103–10.

30. "The flat noses of the Sherpas presented difficulty and a little padding was required in some cases." Mandeville, Hornbein Papers (MSS669), box 31, folder 7, letter, John Cotes to Hornbein, January 11, 1962.

31. For example, see the frustration of Bourdillon that the Sherpa can "only" use closed-circuit oxygen "under supervision": T. D. Bourdillon, "The Use of Oxygen Apparatus by Acclimatized Men," *Proceedings of the Royal Society of London, Series B* 143 (1954): 31.

32. The IBP continued to use the blanket term "Eskimo" through the 1970s.

33. Sarah Pickman, "Dress, Image, and Cultural Encounter in the Heroic Age of Polar Exploration," in *Expedition: Fashion from the Extreme* (New York: Thames & Hudson, 2017), 32. See also Ellen Boucher, "Arctic Mysteries and Imperial Ambitions: The Hunt for Sir John Franklin and the Victorian Culture of Survival," *Journal of Modern History* 90 (2018): 40–75.

34. Michael F. Robinson, *The Coldest Crucible: Arctic Exploration and American Culture* (Chicago: University of Chicago Press, 2006), 69; Efram Sera-Shriar, "Arctic Observers:

Richard King, Monogenism and the Historicisation of Inuit through Travel Narratives." *Studies in History and Philosophy of Science Part C* 51 (2015): 23–31.

35. Robinson, *Coldest Crucible*.
36. Robinson, *Coldest Crucible*, chap. 3, "An Arctic Divided: Isaac Hayes and Charles Hall."
37. See, for example, the many references in Peary, *Secrets of Polar Travel*.
38. Janice Cavell, "Going Native in the North: Reconsidering British Attitudes during the Franklin Search, 1848–1859," *Polar Record* 45 (2009): 26.
39. Raymond Priestley, "Twentieth-Century Man against Antarctica," *Nature* 178 (September 1, 1956): 468.
40. Vilhjalmur Stefansson, *My Life with the Eskimo* (New York: Macmillan, 1912); Stefansson, *The Friendly Arctic: The Story of Five Years in Polar Regions* (New York: Macmillan, 1922).
41. Vilhjalmur Stefansson and US War Department, *Arctic Manual* (Washington, DC: Government Printing Office, 1940), 398.
42. Stefansson, *Arctic Manual*, 437.
43. John Wright, "British Polar Expeditions 1919–39," *Polar Record* 26 (1990): 80.
44. Simon Schaffer, "The Asiatic Enlightenment of British Astronomy," in *The Brokered World: Go-Betweens and Global Intelligence, 1770–1820*, ed. S. Schaffer et al. (Sagamore Beach, MA: Watson Publishing, 2009), 49–104.
45. Alan C. Burton and Otto G. Edholm, *Man in a Cold Environment: Physiological and Pathological Effects of Exposure to Low Temperatures* (London: Edward Arnold, 1955), XI. For an example with specific relevance to extreme physiology, air regulators retrieved from downed German aircraft during the Battle of Britain were provided to American researchers at the Aero Medical Laboratory at Wright Field in 1941. These examples were directly used as the basis for a new model of regulator (the A-12) issued to American airmen early in 1944. Douglas H. Robinson, *The Dangerous Sky: A History of Aviation Medicine* (Henley-on-Thames: Foulis, 1973), 171.
46. K. Anderson, "Science and the Savage: The Linnean Society of New South Wales, 1874–1900," *Cultural Geographies* 5 (1998): 133.
47. David Landy, "Pibloktoq (Hysteria) and Inuit Nutrition: Possible Implication of Hypervitaminosis A," *Social Science & Medicine* 21 (1985): 176.
48. Jenny Mai Handford, "Dog Sledging in the Eighteenth Century: North America and Siberia," *Polar Record* 34 (1998): 238–39.
49. There are Western exceptions, although the fact that "indigenous innovation" had to be explicitly argued *for* in 1969 is telling: Milton M. R. Freeman, "Adaptive Innovation among Recent Eskimo Immigrants in the Eastern Arctic Canada," *Polar Record* 14 (1969): 769–81.
50. Graham Rowley, "Snow-House Building," *Polar Record* 2 (1938): 109.
51. R. DeC. W., "The Snow Huts of the Eskimo," *Bulletin of the American Geographical Society* 37 (1905): 674.
52. Louis Malavielle, "Vacances en igloo sur le Mont-Blanc," *La Montaigne* vii (May 1939): 141–51.
53. "Igloos in the Alps," *Polar Record* 3 (1942): 512–16.
54. "Igloos in the Alps," 512.
55. René Dittert, Gabriel Chevalley, and Raymond Lambert, *Forerunners to Everest*, trans. Malcolm Barnes (London: Hamilton & Co., 1956), 98.
56. Clements R. Markham, *The Lands of Silence* (Cambridge: Cambridge University Press, 1921), 341.
57. Douglas Mawson, *Home of the Blizzard* (London: St. Martin's Press, 1999), chap. 20.

58. Mawson, *Home of the Blizzard*, chap. 20.
59. Ursula Rack, "Felix König and the European Science Community across Enemy Lines during the First World War," *Polar Journal* 4 (2014): 90.
60. William E. Siri and Ann Lage, *William E. Siri: Reflections on the Sierra Club: The Environment and Mountaineering, 1950s–1970s*, Sierra Club History Series (Berkeley: Regional Oral History Office, Bancroft Library, University of California, 1979), 253.
61. Siri and Lage, *William E. Siri*, 253.
62. George Lowe, *From Everest to the South Pole* (New York: St. Martin's Press, 1961), 70.
63. Stefansson, *Arctic Manual*, 1:161–62.
64. Stefansson, *Arctic Manual*, 1:161–62.
65. Stefansson, *Arctic Manual*, 1:166.
66. Stefansson, *Arctic Manual*, 1:165.
67. For much more on tents, see Parsons and Rose, *Invisible on Everest*.
68. Parsons and Rose, *Invisible on Everest*, 34.
69. Parsons and Rose, *Invisible on Everest*, 36.
70. RGS, EE/88, folder 3, Progress reports (spare copies). List of equipment from Everest expedition in the basement of the Royal Geographical Society (as of February, 1955).
71. Mandeville, Pugh Papers (MSS491), box 35, folder 15, letter, presumably Pugh to Hunt, January 23, 1953.
72. J. S. Milledge, "Electrocardiographic Changes at High Altitude," *British Heart Journal* 25 (1963): 291.
73. RGS, EE/47, letter, Shebbeare to Norton, September 28, 1932.
74. RGS, EE/47, letter, Shebbeare to Norton, September 28, 1932.
75. RGS, EE/54/1, Mount Everest Committee Meeting, November 5, 1934. Comments and suggestions—FS Smythe [marked 42/1/3].
76. RGS, EE/33/3/4, Report on the Mount Everest oxygen cylinders & apparatus.
77. Wilfred Noyce, *South Col: The Personal Account of One Man's Adventures on Everest* (London: Reprint Society, 1955), 44.
78. Ryan Johnson, "European Cloth and 'Tropical' Skin: Clothing Material and British Ideas of Health and Hygiene in Tropical Climates," *Bulletin of the History of Medicine* 83 (2009): 530–60.
79. Clare Roche, "Women Climbers 1850–1900: A Challenge to Male Hegemony?" *Sport in History* 3 (2013): 236–59.
80. K. Asahina, "Japanese Antarctic Expedition of 1911–12," in *Polar Human Biology: The Proceedings of the SCAR/IUPS/IUBS Symposium on Human Biology and Medicine in the Antarctic*, ed. O. G. Edholm and E. K. Eric Gunderson (London: Heinemann Medical, 1973), 8–14.
81. L. H. Newburgh, ed., *The Physiology of Heat Regulation and the Science of Clothing; Prepared at the Request of the Division of Medical Sciences, National Research Council* (Philadelphia: W. B. Saunders, 1949).
82. Frederick R. Wulsin, "Adaptations to Climate among Non-European Peoples," in Newburgh, *Physiology of Heat Regulation*, 1–50.
83. Harry Collins, *Changing Order: Replication and Induction in Scientific Practice* (London: Sage Publications, 1985).
84. Siple also writes of "hybrid combinations of Eskimo clothing with *ordinary conventional* cold weather garments [my emphasis]." Paul A. Siple, "Clothing and Climate," in Newburgh, *Physiology of Heat Regulation*, 433–41.
85. G. M. Budd, "Skin Temperature, Thermal Comfort, Sweating, Clothing and Activity of Men Sledging in Antarctica," *Journal of Physiology* 186 (1966): 202.

86. RGS, EE/88, #5, Equipment, Clothing (spare copies).

87. Scott Polar Research Institute [hereafter SPRI], Vivian Fuchs Papers (GB 15), ms1536/2, diary, July 13, 1957.

88. SPRI, Vivian Fuchs Papers (GB 15), ms1536/2, diary, July 13, 1957.

89. Siple, "Clothing and Climate," 433.

90. F. Galton, *The Art of Travel*, 5th ed. (London: John Murray, 1872), 112–13.

91. A. P. Gagge, A. C. Burton, and H. C. Bazett, "A Practical System of Units for the Description of the Heat Exchange of Man with His Environment," *Science* 94 (1941): 428–30.

92. Paul A. Siple and Charles F. Passel, "Measurements of Dry Atmospheric Cooling in Subfreezing Temperatures," *Proceedings of the American Philosophical Society* 89 (1945): 177–99.

93. A. F. Rogers and R. J. Sutherland, *Antarctic Climate, Clothing and Acclimatization: Final Scientific Report* (Bristol: Bristol University Department of Physiology, 1971), section VII.3, section X.5.

94. L. G. C. E. Pugh, "Tolerance to Extreme Cold at Altitude in a Nepalese Pilgrim," *Journal of Applied Physiology* 18 (1963): 1236; Burton and Edholm, *Man in a Cold Environment*, 55.

95. Rogers and Sutherland, *Antarctic Climate*, section X.8.

96. Mountain Heritage Trust, "Mallory Replica Clothing," http://www.mountain-heritage.org/projects/mallory-replica-clothing-2/ (accessed December 2018).

97. Mary B. Rose and Mike Parsons, *Mallory Myths and Mysteries: The Mallory Clothing Replica Project* (Keswick, Cumbria: Mountain Heritage Trust, 2006); George W. Rodway, "Mountain Clothing and Thermoregulation: A Look Back," *Wilderness & Environmental Medicine* 23 (2012): 91–94.

98. G. Hoyland, "Testing Mallory's Clothes on Everest," *Alpine Journal* 112 (2007): 243–46.

99. Paul A. Siple, "General Principles Governing Selection of Clothing for Cold Climates," *Proceedings of the American Philosophical Society* 89 (1945): 200–234.

100. John Giaever, *The White Desert: The Official Account of the Norwegian-British-Swedish Antarctic Expedition* (London: Chatto & Windus, 1954). Giaever, the Norwegian expedition leader, was relatively unusual for the period in being willing to entirely eschew the rhetoric of hardship and manliness on the issue of furs versus other clothing: "It should be purely a matter of personal choice whether a man wishes to be tough and to rough it in the polar regions as everywhere else. Personally I have never had any such ambition; for both in the Arctic and the Antarctic there are all too frequent occasions when it is in any case quite impossible to avoid having a rough time. It is really superfluous to plan hardships deliberately." Giaever, *White Desert*, 21.

101. Carl Murray, "The Use and Abuse of Dogs on Scott's and Amundsen's South Pole Expeditions," *Polar Record* 44 (2008): 303–10.

102. Murray, "Use and Abuse of Dogs." See also John Wright specifically contradicting Shackleton's claim that "the Eskimo" mistreated their dogs: John Wright, "The Polar Eskimos," *Polar Record* 3 (1939): 122.

103. Andrew Croft, "West Greenland Sledge Dogs," *Polar Record* 2 (1937): 77.

104. T. Howard Somervell, *After Everest: The Experiences of a Mountaineer and Medical Missionary*, 2nd ed. (London: Hodder & Stoughton, 1939), 57–58.

105. "The nut food contains cashew-nut cream, peanut cream, soya flour, groundnut oil, 'yeastrel', dehydrated onion, celery, sage, mace and salt." W. R. B. Battle, "An Experimental Concentrated Nut Food," *Polar Record* 7 (1954): 54.

106. L. G. C. E. Pugh, "Physiological and Medical Aspects of the Himalayan Scientific and Mountaineering Expedition," *British Medical Journal* 2 (September 8, 1962): 625.

107. For one example, see Paul Gilchrist, "The Politics of Totemic Sporting Heroes and the Conquest of Everest," *Anthropological Notebooks* 12 (2006): 41.

108. Mandeville, Pugh Papers (MSS491), box 35, folder 1, letter, M. W. Grant to Pugh, November 14, 1952.

109. Iain A. McCormick et al., "A Psychometric Study of Stress and Coping during the International Biomedical Expedition to the Antarctic (IBEA)," *Journal of Human Stress* 11 (1985): 153.

110. Anthony J. W. Taylor and Iain A. McCormick, "Human Experimentation during the International Biomedical Expedition to the Antarctic (IBEA)," *Journal of Human Stress* 11 (1985): 161–62.

111. Mandeville, Pugh Papers (MSS491), box 39, folder 5, High-Altitude Ration Questionnaire c. 1952.

112. Noel Barber, *The White Desert* (London: Hodder & Stoughton, 1958), 106.

113. Deborah Neill, "Finding the 'Ideal Diet': Nutrition, Culture, and Dietary Practices in France and French Equatorial Africa, c. 1890s to 1920s," *Food and Foodways* 17 (2009): 15.

114. Heggie, "Rationalised and Rationed."

115. E. F. Adolph, ed., *Physiology of Man in the Desert* (New York: Interscience Publishers, 1947); see, in particular, chapter 6, "Urinary Excretion of Water and Solutes."

116. Indeed, this diet was once used as an explanation for the name "Eskimo" and for its perjorativeness—it was thought to be a corruption of terms meaning "raw meat eater." This origin is debated by linguists.

117. Kåre Rodahl and T. Moore, "The Vitamin A Content and Toxicity of Bear and Seal Liver," *Biochemical Journal* 37 (1943): 166–68.

118. Kåre Rodahl, *Between Two Worlds: A Doctor's Log-Book of Life amongst the Alaskan Eskimos*, 2nd ed. (London: Heinemann, 1964), chaps. 9, 10.

119. Rodahl, *Between Two Worlds*, 134.

120. Vilhjalmur Stefansson, *The Fat of the Land* (enlarged edition of *Not by Bread Alone*) (New York: Macmillan, 1960), 74.

121. Stefansson, *Fat of the Land*, 74.

122. Hugh Ruttledge, *Everest: The Unfinished Adventure* (London: Hodder & Stoughton, 1937); T. S. Blakeney, "Maurice Wilson and Everest, 1934," *Alpine Journal* 70 (1965): 269–72.

123. [Frank Debenham?], "The Eskimo Kayak," *Polar Record* 1 (1934): 54.

124. [Debenham?], "The Eskimo Kayak," 57.

125. Mandeville, Pugh Papers (MSS491), box 6, folder 25, letter, Brotherhood to Grahame, October 10, 1972.

126. Rowley, "Snow-House Building," 109.

127. "About Pemmican," Classic Jerky Company, http://pemmican.com/about-pemmican/ (accessed May 2017).

128. Elena Aronova, Karen S. Baker, and Naomi Oreskes, "Big Science and Big Data in Biology: From the International Geophysical Year through the International Biological Program to the Long Term Ecological Research (LTER) Network, 1957–Present," *Historical Studies in the Natural Sciences* 40 (2010): 183–224.

129. Joanna Radin, *Life on Ice: A History of New Uses for Cold Blood* (Chicago: University of Chicago Press, 2017). On "vanishing" populations, see also Soraya de Chadarevian,

"Human Population Studies and the World Health Organization," *Dynamis* 35 (2015): 359–88.

130. Joanna Radin, "Latent Life: Concepts and Practices of Human Tissue Preservation in the International Biological Program," *Social Studies of Science* 43 (2013): 484–508.

131. Henrik Forsius, Aldur W. Eriksson, and Johan Fellman, "The International Biological Program/Human Adaptability Studies among the Skolt Sami in Finland (1966–1970)," *International Journal of Circumpolar Health* 71 (2012): 1–5.

CHAPTER FIVE

1. John B. West, *High Life: A History of High-Altitude Physiology and Medicine* (New York: Published for the American Physiological Society by Oxford University Press, 1998), 240; George W. Rodway, "Ulrich C. Luft and Physiology on Nanga Parbat: The Winds of War," *High Altitude Medicine & Biology* 10 (2009): 89–96.

2. West, *High Life*, 250.

3. Paul Bauer, "Nanga Parbat, 1937," *Himalayan Journal* 10 (1938), https:///www.himalayanclub.org/jnl/10/9/nanga-parbat-1937/.

4. Bauer had led three teams that failed to reach the summit of Kangchenjunga and had subsequently failed to raise funds for further expeditions to that mountain, as interest had switched to Nanga Parbat.

5. Bauer, "Nanga Parbat."

6. "Lagen sie friedlich in ihre Zelte, ihre Gesichter zeigten keine Spur einer Angst vor de nahendne Unheil." University of California, San Diego, Mandeville Special Collections [hereafter Mandeville], Ulrich Cameron Luft Papers (MSS475), box 40, folder 8, manuscript, "Ganz Deutschland hielt den Atem an . . . ," Munich, November 7, 1937.

7. Bauer, "Nanga Parbat"; Harald Hoebusch, "Ascent into Darkness: German Himalaya Expeditions and the National Socialist Quest for High-Altitude Flight," *International Journal of the History of Sport* 24 (2007): 526.

8. Aviation technology was of sustained strategic interest to the US military; see, for example, the retrieved information in United States Air Force, *German Aviation Medicine, World War II*, vol. 2 (Washington, DC: Government Printing Office, 1950); A. Jacobson, *Operation Paperclip: The Secret Intelligence Program that Brought Nazi Scientists to America* (Boston: Little, Brown, 2014).

9. Rodway, "Ulrich C. Luft."

10. Mandeville, Pugh Papers (MSS491), box 42, folder 8, letter, Nevison to "Griff," November 13, 1961.

11. Rodway, "Ulrich C. Luft"; see also Karl Heinz Roth, "Flying Bodies—Enforcing States: German Aviation Medical Research from 1925 to 1975 and the Deutsche Forschungsgemeinschaft," in *Man, Medicine, and the State: The Human Body as an Object of Government Sponsored Medical Research in the 20th Century*, ed. Wolfgang U. Eckart (Stuttgart: Franz Steiner Verlag Wiesbaden GmbH, 2006), 108–32; Jonathan D. Moreno, *Undue Risk: Secret State Experiments on Humans* (New York: W. H. Freeman, 1999); in particular, chapter 4: "Deals with Devils."

12. F. Viault, "Sur l'augmentation considérable de nombre des globules rouges dans le sang chez les habitants des hautes plataux de l'Amérique du Sud," *Comptes rendues de l'Académie des Sciences* 111 (1890): 917–18. See also West, *High Life*, 200–201. West notes that Viault's increases, if accurate, are much higher than modern studies would expect, and are therefore probably due to significant dehydration as well as increased red blood cell production.

13. Mandeville, Pugh Papers (MSS491), box 39, folder 1, letter, J. S. Horn to Edholm, June 22, 1953.

14. Mandeville, Pugh Papers (MSS491), box 39, folder 1, letter, Pugh to J. S. Horn, July 14, 1953.

15. W. Schneider, "Blood Transfusion in Peace and War, 1900–1918," *Social History of Medicine* 10 (1997): 105–26.

16. Mandeville, Hornbein Papers (MSS669), box 32, folder 5, letter, Tom to Ross Paul, November 21, 1963.

17. R. A. Zink et al., "Hemodilution: Practical Experiences in High Altitude Expeditions," in *High Altitude Physiology and Medicine*, ed. W. Brendel and R. A. Zink (New York: Springer-Verlag, 1982), 291–97.

18. Mandeville, West Papers (MSS444), box 81, folder 6, Research proposal by Frank H. Sarnquist to American Lung Association (successful), May 31, 1982.

19. Mandeville, West Papers (MSS444), box 81, folder 6, letter, ?West to AMREE team, November 17, 1980.

20. Hugo Chiodi, "Respiratory Adaptations to Chronic High Altitude Hypoxia," *Journal of Applied Physiology* 10 (1957): 81–87; F. Kreuzer and Z. Turek, "Influence of the Position of the Oxygen Dissociation Curve on the Oxygen Supply to Tissues," in *High Altitude Physiology and Medicine*, ed. W. Brendel and R. A. Zink (New York: Springer-Verlag, 1982), 66–72.

21. Zink et al., "Hemodilution," 295; José Luis Bermúdez, "Climbing on Kangchenjunga since 1955," *Alpine Journal* 101 (1996): 50–56.

22. F. H. Sarnquist, R. Schoene, and P. Hackett, "Exercise Tolerance and Cerebral Function after Acute Hemodilution of Polycythemic Mountain Climbers," *Physiologist* 25 (1982): 327.

23. Zink et al., "Hemodilution," 296.

24. K. Messmer, "Oxygen Transport Capacity," and on the Austrian expedition, Oswald Oelz, "How to Stay Healthy While Climbing Mount Everest," both in *High Altitude Physiology and Medicine*, ed. W. Brendel and R. A. Zink (New York: Springer-Verlag, 1982), 21–27, 298–300, respectively.

25. T. Howard Somervell, "Note on the Composition of Alveolar Air at Extreme Heights," *Journal of Physiology* 60 (1925): 282–85.

26. Mandeville, Pugh Papers (MSS491), box 39, folder 1, letter, J. S. Horn to Edholm, June 22, 1953; this work was eventually published as a short note: L. G. C. E. Pugh, "Haemoglobin Levels on the British Himalayan expeditions to Cho Oyu in 1952 and Everest in 1953," *Journal of Physiology* 126-Supp. (1964): 38–39.

27. John B. Winslow, "High-Altitude Polycythemia," in *High Altitude and Man*, ed. John B. West and Sukhamay Lahiri (Bethesda, MD: American Physiological Society, 1984), 163–73.

28. Nancy Stepan, *The Idea of Race in Science: Great Britain 1800–1960* (Houndmills, Hampshire: Macmillan, 1987); see, in particular, chapter 6, "A Period of Doubt: Race Science before the Second World War," and chapter 7, "After the War: New Science & Old Controversies."

29. For a review of these for the Sherpa, see Sherry B. Ortner, *Life and Death on Mt. Everest: Sherpas and Himalayan Mountaineering* (Princeton, NJ: Princeton University Press, 1999).

30. Joseph Barcroft, *The Respiratory Function of the Blood, Part I: Lessons from High Altitudes* (Cambridge: Cambridge University Press, 1925), 176.

31. Eric T. Jennings, *Curing the Colonizers: Hydrotherapy, Climatology, and French Colonial Spas* (Durham, NC: Duke University Press, 2006), 75–78; Dane Kennedy, *The Magic Mountains: Hill Stations and the British Raj* (Berkeley: University of California Press, 1996); J. T. Kenny, "Climate, Race, and Imperial Authority: The Symbolic Landscape of the British Hill Station in India," *Annals of the Association of American Geographers* 85 (1995): 694–714; Kenny, "Claiming the High Ground: Theories of Imperial Authority and the British Hill Stations in India," *Political Geography* 16 (1997): 655–73.

32. Ana Cecilia Rodríguez de Romo and José Rogelio Pérez Padillia, "The Mexican Response to High Altitudes in the 1890s: The Case of a Physician and his 'Magic Mountain,'" *Medical History* 42 (2003): 493–516.

33. H. P. Lobenhoffer, R. A. Zink, and W. Brendel, "High Altitude Pulmonary Edema: Analysis of 166 Cases," in *High Altitude Physiology and Medicine*, ed. W. Brendel and R. A. Zink (New York: Springer-Verlag, 1982), 219–31.

34. Carlos Monge Medrano, *Acclimatization in the Andes* (Baltimore: Johns Hopkins University Press, 1948), xii.

35. Monge Medrano, *Acclimatization*, xii.

36. Monge Medrano, *Acclimatization*, 47.

37. Monge Medrano, *Acclimatization*, ix.

38. Cf. John B. West, "Barcroft's Bold Assertion: All Dwellers at High Altitudes Are Persons of Impaired Physical and Mental Powers," *Journal of Physiology* 594 (2016): 1127–34.

39. Monge Medrano, *Acclimatization*, 62.

40. Ortner, *Life and Death*; Ortner, "Thick Resistance: Death and the Cultural Construction of Agency in Himalayan Mountaineering," *Representations* 59 (1997): 135–62. Pain has long been recognized by historians as a medical concept deeply shaped by racialized ideas: Joanna Bourke, "Pain Sensitivity: An Unnatural History from 1800 to 1965," *Journal of Medical Humanities* 35 (2014): 301–19.

41. Lobenhoffer, Zink, and Brendel, "High Altitude Pulmonary Edema."

42. L. G. C. E. Pugh, "Tolerance to Extreme Cold at Altitude in a Nepalese Pilgrim," *Journal of Applied Physiology* 18 (1963): 1234–38.

43. J. G. Wilson, "The Himalayan Schoolhouse Expeditions," *Alpine Journal* 70 (1965): 226–39.

44. S. Lahiri and J. S. Milledge, "Sherpa Physiology," *Nature* 207 (1965): 611.

45. S. Lahiri and J. S. Milledge, "Acid-Base in Sherpa Altitude Residents and Lowlanders at 4880 m," *Respiration Physiology* 2 (1967): 332.

46. Lahiri and Milledge, "Sherpa Physiology," 612.

47. Peter H. Hackett et al., "Control of Breathing in Sherpas at Low and High Altitude," *Journal of Applied Physiology* 49 (1980): 374.

48. Frederic Jackson, "The Heart at High Altitude," *British Heart Journal* 30 (1968): 292.

49. Mandeville, Pugh Papers (MSS491), box 42, folder 10, letter, Elsner to Pugh, June 27, 1960.

50. Jeremy Windsor and George W. Rodway, "Heights and Haematology: The Story of Haemoglobin at Altitude," *Postgraduate Medical Journal* 83 (2007): 148–51.

51. Mandeville, Pugh Papers (MSS491), box 10, folder 35, letter, G. Pugh to P. O. Williams [MRC], November 25, 1958.

52. Frederic Jackson and Hywel Davies, "The Electrocardiogram of the Mountaineer at High Altitude," *British Heart Journal* 22 (1960): 671–85.

53. J. H. Emlyn Jones, "Ama Dablam, 1959," *Alpine Journal* 65 (1960): 1–10.

54. The classics on this topic are Londa Schiebinger, "Why Mammals Are Called Mammals: Gender Politics in Eighteenth-Century Natural History," *American Historical*

Review 98 (1993): 382–411; and Emily Martin, "The Egg and the Sperm: How Science Has Constructed a Romance Based on Stereotypical Male-Female Roles," *Signs* 16 (1991): 485–501.

55. Mandeville, Pugh Papers (MSS491), box 35, folder 13, diary, April 11, 1953.

56. Martina Gugglberger, "Climbing Beyond the Summits: Social and Global Aspects of Women's Expeditions in the Himalayas," *International Journal of the History of Sport* 32 (2015): 597–613.

57. Gugglberger, "Climbing Beyond," 605, 609.

58. James A. Horscroft et al., "Metabolic Basis to Sherpa Altitude Adaptation," *Proceedings of the National Academy of Sciences* 114 (2017): 6382–87.

59. Warwick Anderson, "Hybridity, Race, and Science: The Voyage of the *Zaca*, 1934–1935," *Isis* 103 (2012): 229–53.

60. In fact, Kestner's initial research was on dogs. O. Kestner, "Klimatologische Studien. L Der wirksame Anteil des Hoheriklimas," *Zeitschrift für Biologie* 73 (1921): 1–6; as cited in H. H. Mitchell and Marjorie Edman, *Nutrition and Resistance to Climatic Stress with Particular Reference to Man*, Report: Quartermaster Food and Container Institute for the Armed Forces, Research and Development Branch (November 1949), 41, www.dtic.mil/dtic/tr/fulltext/u2/a581922.pdf (accessed July 2018).

61. D. V. Latham and C. Gillman, "Kilimanjaro and Some Observations on the Physiology of High Altitudes in the Tropics," *Geographical Journal* 68 (1926): 492.

62. E. M. Glaser, "Acclimatization to Heat and Cold," *Journal of Physiology* 110 (1949): 335.

63. Glaser, "Acclimatization to Heat and Cold."

64. E. F. Adolph, ed., *Physiology of Man in the Desert* (New York: Interscience Publishers, 1947). 17.

65. University of Adelaide Rare Books & Special Collections [hereafter Adelaide], W. V. Macfarlane Papers 1947–1985 (MS0006), F2/51 US Army Medical Research Command—Water and electrolyte metabolism of desert Aboriginals and New Guinea Melanesians.

66. R. H. Fox et al., "A Study of Temperature Regulation in New Guinea People," *Philosophical Transactions of the Royal Society of London, Series B* 268 (1974): 375–91.

67. A. Grenfell Price, *White Settlers in the Tropics* (New York: American Geographical Society, 1939).

68. Warwick Anderson, *The Cultivation of Whiteness: Science, Health, and Racial Destiny in Australia* (Durham, NC: Duke University Press, 2006).

69. Adelaide, W. V. Macfarlane Papers 1947–1985 (MS0006), F2/37, letter, Macfarlane to Law, March 11, 1980.

70. Adelaide, W. V. Macfarlane Papers 1947–1985 (MS0006), F2/1, letter, Macfarlane to Kirk, May 21, 1969.

71. Adelaide, W. V. Macfarlane Papers 1947–1985 (MS0006), F2/6, Macfarlane to Mr. R. Ballantyne, Building Research Station, CSIRO, Victoria, January 18, 1978.

72. Jean Rivolier, *Man in the Antarctic: The Scientific Work of the International Biomedical Expedition to the Antarctic (IBEA)* (London: Taylor & Francis, 1988), 105.

73. Rivolier, *Man in the Antarctic*, 105.

74. Joel B. Hagen, "Bergmann's Rule, Adaptation, and Thermoregulation in Arctic Animals: Conflicting Perspectives from Physiology, Evolutionary Biology, and Physical Anthropology after World War II," *Journal of the History of Biology* 50 (2017): 235–65.

75. "Davenport also demonstrates that some of the mulattoes have unexpected combinations of long legs and short bodies, or long bodies and short legs. Other individuals have the long legs of the negro and the short arms of the white, which would

put them at a disadvantage in picking up things from the ground." C. B. Davenport and Morris Steggerda, "Race Crossing in Jamaica" (Washington: Carnegie Institution, 1929), quoted in Grenfell Price, *White Settlers in the Tropics*, 180.

76. Hagen, "Bergmann's Rule," 248.

77. Hagen, "Bergmann's Rule". We should acknowledge that some of this counterargument relied on the belief that adaptive mechanisms seen in non-European populations were at least "latent" in people of European descent; these mechanisms included a generalized acclimatization to cold, which, of course, many other physiologists were denying existed in any population.

78. Frederick A. Itoh, "Physiology of Circumpolar People," in *The Human Biology of Circumpolar Populations*, ed. Frederick A. Milan (Cambridge: Cambridge University Press, 1980), 288.

79. Mandeville, Pugh Papers (MSS491), box 7, folder 31, letter, Dr. Raymond Greene to Pugh, July 11, 1957.

80. R. F. Hellon et al., "Natural and Artificial Acclimatisation to Hot Environments," *Journal of Physiology* 132 (1956): 559–76.

81. Fox et al., "Study of Temperature Regulation," 390.

82. D. C. Wilkins, "Heat Acclimatization in the Antarctic," *Journal of Physiology* 214 (1971): 15–16.

83. Steven M. Horvath and Elizabeth C. Horvath, *The Harvard Fatigue Laboratory: Its History and Contributions*, International Research Monograph Series in Physical Education (Englewood Cliffs, NJ: Prentice-Hall, 1973), 147.

84. Javier Arias-Stella, "Morphological Patterns: Mechanism of Pulmonary Arterial Hypertension," in *Life at High Altitudes* [Proceedings of the Special Session Held during the Fifth Meeting of the PAHO Advisory Committee on Medical Research, June 15, 1966], ed. A. Hurtado (Washington, DC: Pan American Health Organisation/World Health Organisation, 1966), 11.

85. D. A. Giussani et al., "Hypoxia, Fetal and Neonatal Physiology: 100 Years On from Sir Joseph Barcroft: Editorial," *Journal of Physiology* 594 (2016): 1107.

86. A. Hurtado, "Natural Acclimatization to High Altitudes: Review of Concepts," in Hurtado, *Life at High Altitudes*, 7.

87. "Polar Medicine," *Lancet* 274 (November 7, 1959): 787.

88. Nea Morin, *A Woman's Reach: Mountaineering Memoirs* (London: Eyre & Spottiswoode, 1968), 215.

89. Twentieth-century writers sometimes use the single word "Ama" to refer to both Japanese and Korean diving women.

90. Hermann Rahn, "Lessons from Breath Holding," in *The Regulation of Human Respiration: The Proceedings of the J. S. Haldane Centenary Symposium Held in the University Laboratory of Physiology, Oxford*, ed. D. J. Cunningham and B. B. Lloyd (Oxford: Blackwell Scientific, 1963), 293–302.

91. Suk Ki Hong and Hermann Rahn, "The Diving Women of Korea and Japan," in *Human Physiology and the Environment in Health and Disease: Readings from Scientific American*, ed. Arthur J. Vander (San Francisco: W. H. Freeman, 1976), chap. 9, 92–101.

92. Hong and Rahn, "Diving Women," 100.

93. Hong and Rahn, "Diving Women," 101.

94. Vilhjalmur Stefansson and US War Department, *Arctic Manual* (Washington, DC: Government Printing Office, 1940), 293.

95. H. E. Lewis and J. P. Masterton. "British North Greenland Expedition 1952–54: Medical and Physiological Aspects [Part 1]," *Lancet* 266 (September 3, 1955): 499.

96. G. M. Budd and N. Warhaft, "Cardiovascular and Metabolic Responses to Noradrenaline in Man, before and after Acclimatization to Cold in Antarctica," *Journal of Physiology* 186 (1966): 233–42.

97. Mandeville, Pugh Papers (MSS491), box 7, folder 28, letter, Goldberger to Leathes, December 2, 1970.

98. Vanessa Heggie, "'Only the British Appear to Be Making a Fuss': The Science of Success and the Myth of Amateurism at the Mexico Olympiad, 1968," *Sport in History* 28 (2008): 213–35; Alison M. Wrynn, "'A Debt Was Paid Off in Tears': Science, IOC Politics and the Debate about High Altitude in the 1968 Mexico City Olympics," *International Journal of the History of Sport* 23 (2006): 1152–72.

99. Mandeville, Hornbein Papers (MSS669), box 32, folder 5, letter, Tom to Ross Paul, November 21, 1963.

100. John Gleaves, "Manufactured Dope: How the 1984 US Olympic Cycling Team Rewrote the Rules on Drugs in Sports," *International Journal of the History of Sport* 32 (2015): 89–107.

101. Vanessa Heggie, *A History of British Sports Medicine* (Manchester: Manchester University Press, 2011).

102. This is a literature too large to cite fully, but good starting points are Paul Dimeo, *A History of Drug Use in Sport, 1876–1976: Beyond Good and Evil* (New York: Routledge/Taylor & Francis, 2007); Rob Beamish and Ian Ritchie, "From Fixed Capacities to Performance-Enhancement: The Paradigm Shift in the Science of 'Training' and the Use of Performance-Enhancing Substances," *Sport in History* 25 (2005): 412–33; John M. Hoberman, *Mortal Engines: The Science of Performance and the Dehumanization of Sport* (Caldwell, NJ: Blackburn Press, 2001).

CHAPTER SIX

1. L. G. C. E. Pugh, "Clothing Insulation and Accidental Hypothermia in Youth," *Nature* 209 (1966): 1281–86.

2. L. G. C. E. Pugh, "Cold Stress and Muscular Exercise, with Special Reference to Accidental Hypothermia," *British Medical Journal* 2 (May 6, 1967): 333–37; Pugh, "Clothing Insulation," 1286.

3. L. G. C. E. Pugh, "Accidental Hypothermia in Walkers, Climbers, and Campers: Report to the Medical Commission on Accident Prevention," *British Medical Journal* 1 (January 15, 1966).

4. Pugh, "Accidental Hypothermia," 123.

5. Pugh, "Clothing Insulation."

6. Mark Jackson, *The Age of Stress: Science and the Search for Stability* (Oxford: Oxford University Press, 2013); Anson Rabinbach, *The Human Motor: Energy, Fatigue and the Origins of Modernity* (Berkeley: University of California Press, 1992); Anna Katharina Schaffner, *Exhaustion: A History* (New York: Columbia University Press, 2016); Robin Wolfe Scheffler, "The Fate of a Progressive Science: The Harvard Fatigue Laboratory, Athletes, the Science of Work and the Politics of Reform," *Endeavour* 35 (June 2011): 48–54.

7. These changes are outlined in Jackson, *Age of Stress*, although, as Layne Karafantis has noted, there are still lacunae in work on *field* psychology. Karafantis, "Sealab II and Skylab: Psychological Fieldwork in Extreme Spaces," *Historical Studies in the Natural Sciences* 43 (2013): 551–88.

8. And, of course, in choosing future polar and high-altitude personnel; psychological testing had been part of the recruitment process for American Antarctic personnel

since the early 1960s. E. K. Eric Gunderson, "Psychological Studies in Antarctica," in *Human Adaptability to Antarctic Conditions,* ed. E. K. Eric Gunderson, Antarctic Research Series, vol. 22 (Washington, DC: American Geophysical Union, 1974), 115–31. See also Philip W. Clements, *Science in an Extreme Environment: The 1963 American Mount Everest Expedition* (Pittsburgh: University of Pittsburgh Press, 2018).

9. E. M. Glaser, "Immersion and Survival in Cold Water," *Nature* 166 (1950): 1068.

10. "The metabolic cost of sustained swimming . . . is not greatly in excess of the heat production from shivering, at any rate in peak periods. Heat loss from a moving body as compared with a subject keeping still would possibly be increased, and for thin subjects keeping still might be better than swimming. The evidence on this point is not yet adequate." Alan C. Burton and Otto G. Edholm, eds., *Man in a Cold Environment. Physiological and Pathological Effects of Exposure to Low Temperatures* (London: Edward Arnold, 1955), 212.

11. L. G. C. E. Pugh and O. G. Edholm, "The Physiology of Channel Swimmers," *Lancet* 269 (1955): 761–68; "Physiology of Channel Swimmers," *British Medical Journal* 2 (September 3, 1960): 725; Vanessa Heggie, *A History of British Sports Medicine* (Manchester: Manchester University Press, 2011).

12. Steven M. Horvath and Elizabeth C. Horvath, *The Harvard Fatigue Laboratory: Its History and Contributions,* International Research Monograph Series in Physical Education (Englewood Cliffs, NJ: Prentice-Hall, 1973), 155. On the links between the ethics of this work and Nazi experiments, see Paul Weindling, *John W. Thompson: Psychiatrist in the Shadow of the Holocaust* (Rochester, NY: University of Rochester Press, 2010), esp. 138–51.

13. Burton and Edholm, *Man in a Cold Environment,* 212.

14. O. G. Edholm and A. L. Bacharach, eds., *Exploration Medicine: Being a Practical Guide for Those Going on Expeditions* (Bristol: John Wright & Sons, 1965).

15. R. L. Berger, "Nazi Science—the Dachau Hypothermia Experiments," *New England Journal of Medicine* 322 (1990): 1435–40.

16. Wilfred Noyce, *South Col: One Man's Adventure on the Ascent of Everest 1953* (London: Reprint Society, 1955), 198.

17. J. S. Haldane, A. M. Kellas, and E. L. Kennaway, "Experiments on Acclimatisation to Reduced Atmospheric Pressure," *Journal of Physiology* 53 (1919): 181–206; J. M. H. Campbell, C. Gordon Douglas, and F. G. Hobson, "The Respiratory Exchange of Man during and after Muscular Exercise," *Philosophical Transactions of the Royal Society of London, Series B* 210 (1921): 1–47.

18. Ove Wilson, "Physiological Changes in Blood in the Antarctic," *British Medical Journal* 2 (December 26, 1953): 1425.

19. C. B. Warren, "The Medical and Physiological Aspects of the Mount Everest Expeditions," *Geographical Journal* 90 (1937): 136.

20. As examples: Royal Geographical Society Archives [hereafter RGS], EE/45, letter, Ruttledge to Greene, November 22, 1932; EE/66.11, letter, W. G. Lowe to Dr. B. N. Wallies, October 15, 1953.

21. University of California, San Diego, Mandeville Special Collections [hereafter Mandeville], Pugh Papers (MSS491), box 7, folder 50.

22. Mandeville, Pugh Papers (MSS491), box 7, folder 50, letter, Hyman to Pugh, February 1967.

23. W. R. Keatinge, "The Effect of Repeated Daily Exposure to Cold and of Improved Physical Fitness on the Metabolic and Vascular Response to Cold Air," *Journal of Physiology* 157 (1961): 209.

24. Mandeville, Pugh Papers (MSS491), box 42, folder 10, letter, Pugh to "Ed," April 6, 1959.

25. Clements, *Science in an Extreme Environment*.

26. Shirley V. Scott, "Ingenious and Innocuous? Article IV of the Antarctic Treaty as Imperialism," *Polar Journal* 1 (June 2011): 51–62.

27. Other funders included the American Alpine Club, the National Geographic Society, the Parisian Servier Laboratories, the Explorers Club, and private donations. F. H. Sarnquist, "Physicians on Mount Everest," *Western Journal of Medicine* 139 (1983): 480–85.

28. On the TAE: "During this stage of the journey Geoffrey Pratt was found to be suffering from severe cumulative carbon monoxide poisoning and had to be given oxygen from the welding equipment." Vivian Fuchs, "The Commonwealth Trans-Antarctic Expedition," *Geographical Journal* 124 (1958): 447. And at Silver Hut: "When we were measuring carbon monoxide diffusion capacities in the Silver Hut in 1961, I found on one occasion that the expired carbon monoxide concentration exceeded the inspired concentration (which was about 0.05%). We traced the problem to the fact that a Sherpa had been using a primus stove inside the Silver hut which was well sealed. I think that 0.05% is equivalent to 500 parts per million so the expired concentration exceeded that. We must have had a lot of carbon monoxide on board." Mandeville, West Papers (MSS444), box 8, folder 4, letter, West to Hackett, December 5, 1988.

29. "Such a reaction is extremely rare, but it brought home to us that an invasive procedure in such a remote environment can easily turn into a serious unpleasant situation because there was little if any way to treat him. Fortunately the reaction was short-lived." John B. West, *Everest: The Testing Place* (New York: McGraw Hill, 1985), 146.

30. Charles S. Houston, "Introductory Address: Lessons to Be Learned from High Altitude," *Postgraduate Medical Journal* 55 (July 1979): 450.

31. Houston, "Introductory Address," 450.

32. John Hunt, *The Ascent of Everest* (London: Hodder & Stoughton, 1953), 71.

33. Hunt, *Ascent of Everest*, 71.

34. A. F. Rogers and R. J. Sutherland, *Antarctic Climate, Clothing and Acclimatization: Final Scientific Report* (Bristol: Bristol University Department of Physiology, 1971), section II.2.

35. Charles J. Eagan, "Resistance to Finger Cooling Related to Physical Fitness," *Nature* 200 (1963): 852.

36. Workers were "sampled" between five and eleven times over the course of the year. D. L. Easty, A. Antonis, and I. Bersohn, "Adipose Fat Composition in Young Men in Antarctica," *British Antarctic Survey Bulletin* 13 (1967): 41–45.

37. H. E. Lewis and J. P. Masterton, "British North Greenland Expedition 1952–54: Medical and Physiological Aspects [Part 2]," *Lancet* 266 (1955): 552.

38. Lewis and Masterton, "British North Greenland [Part 2]," 552.

39. Anthony J. W. Taylor and Iain A. McCormick, "Human Experimentation during the International Biomedical Expedition to the Antarctic (IBEA)," *Journal of Human Stress* 11 (1985): 162.

40. J. R. Brotherhood, "The British Antarctic Environment with Special Reference to Energy Expenditure," *Forsvarsmedicine* 8 (1972): 112.

41. "Only one Sherpa learnt to work satisfactorily on the bicycle ergometer at the Silver Hut." L. G. C. E. Pugh, "Physiological and Medical Aspects of the Himalayan

Scientific and Mountaineering Expedition," *British Medical Journal* 2 (September 8, 1962): 624.

42. P. Cerretelli, A. Veicsteinas, and C Marconi, "Anaerobic Metabolism at High Altitude: The Lactacid Mechanism," in *High Altitude Physiology and Medicine*, ed. W. Brendel and R. A. Zink (New York: Springer-Verlag, 1982), 95.

43. Roy J. Shephard, "Work Physiology and Activity Patterns of Circumpolar Eskimos and Ainu: A Synthesis of IBP Data," *Human Biology* 46 (1974): 263–94.

44. A. V. Hill, *Muscular Movement in Man: The Factors Governing Speed and Recovery from Fatigue* (New York: McGraw-Hill, 1927), 3.

45. For more biographical detail, see Harriet Tuckey, *Everest—The First Ascent: The Untold Story of Griffith Pugh, the Man Who Made It Possible* (London: Rider, 2013).

46. Mandeville, West Papers (MSS444), box 25, folder 32, letter, West to Sheldon Shultz.

47. A. H. Brown, "Water Requirements of Man in the Desert," in *Physiology of Man in the Desert*, ed. E. F. Adolph (New York: Interscience Publishers, 1947), 122.

48. Adolph, *Physiology of Man*.

49. The relative advantage of high skin melanin in warmer climates was still a matter of debate in the 1950s: M. L. Thomson, "The Cause of Changes in Sweating Rate after Ultraviolet Radiation," *Journal of Physiology* 112 (1951): 31–42.

50. Joanna Radin, *Life on Ice: A History of New Uses for Cold Blood* (Chicago: University of Chicago Press, 2017). See, in particular, chap. 3, "'Before It's Too Late': Life from the Past."

51. Sherry B. Ortner, "Thick Resistance: Death and the Cultural Construction of Agency in Himalayan Mountaineering," *Representations* 59 (1997): 135–62; Ortner, *Life and Death on Mt. Everest: Sherpas and Himalayan Mountaineering* (Princeton, NJ: Princeton University Press, 1999).

52. Richard Gillespie, "Industrial Fatigue and the Discipline of Physiology," in *Physiology in the American Context, 1850–1940*, ed. Gerald L. Geison (Bethesda, MD: American Physiological Society, 1987), 237–62.

53. Scheffler, "Fate of a Progressive Science"; Carleton B. Chapman, "The Long Reach of Harvard's Fatigue Laboratory, 1926–1947," *Perspectives in Biology and Medicine* 34 (1990): 17–33; Horvath and Horvath, *Harvard Fatigue Laboratory*.

54. Gillespie, "Industrial Fatigue."

55. "The flat noses of the Sherpas presented difficulty and a little padding was required in some cases." Mandeville, Hornbein Papers (MSS669), box 32, folder 7, letter, John Cotes to Tom "Dr. Hornbein," January 11, 1962.

56. Ken Wilson and Mike Pearson, "Post-Mortem of an International Expedition," *Himalayan Journal* 31 (1971): 33–83.

57. Walt Unsworth, *Everest: The Mountaineering History*, 3rd ed. (London: Bâton Wicks, 2000), 409.

58. Hugh Ruttledge, *Everest: The Unfinished Adventure* (London: Hodder & Stoughton, 1937), 44.

59. W. Larry Kenney, David W. DeGroot, and Lacy Alexander Holowatz, "Extremes of Human Heat Tolerance: Life at the Precipice of Thermoregulatory Failure," *Journal of Thermal Biology* 29 (2004): 481.

60. F. P. Ellis et al., "The Upper Limits of Tolerance of Environmental Stress," in *Physiological Responses to Hot Environments*, ed. R. K. Macpherson (London: HMSO, 1960), 158–79.

61. Ortner, *Life and Death*, 288. Schmatz's husband eventually paid for her body to be removed although two men (including one Sherpa porter) died in the attempt.

62. W. Ambach et al., "Corpses Released from Glacier Ice: Glaciological and Forensic Aspects," *Journal of Wilderness Medicine* 3 (1992): 372–76.

63. Unsworth, *Everest*, 409.

64. John V. Pickstone, *Ways of Knowing: A New History of Science, Technology and Medicine* (Chicago: University of Chicago Press, 2001).

65. Proving equipment by taking it out into the field is by no means a modern phenomenon. Nicky Reeves, "'To Demonstrate the Exactness of the Instrument': Mountainside Trials of Precision in Scotland, 1774," *Science in Context* 22 (2009): 323–40.

66. Dorinda Outram, "On Being Perseus: New Knowledge, Dislocation, and Enlightenment Exploration," in *Geography and Enlightenment*, ed. David N. Livingstone and Charles W. J. Withers (Chicago: University of Chicago Press, 1997), 281–94.

67. Naomi Oreskes, "Objectivity or Heroism? On the Invisibility of Women in Science," *Osiris* 11 (1996): 87–113.

68. Clements, *Science in an Extreme Environment.*

69. Outram, "On Being Perseus."

70. On mountains: Clements, *Science in an Extreme Environment.* In Antarctica: Jean Rivolier, "Physiological and Psychological Studies Conducted by Continental European and Japanese Expeditions," in *Human Adaptability to Antarctic Conditions*, ed. E. K. Eric Gunderson, Antarctic Research Series, vol. 22 (Washington, DC: American Geophysical Union, 1974), 63.

BIBLIOGRAPHY

ARCHIVAL RESOURCES

Royal Geographical Society Archives, London, UK

Everest Expeditions (EE/1 to EE/99)

Scott Polar Research Institute, Cambridge, UK

Vivian Fuchs Papers (GB 15)

Mandeville Special Collections, University of California, San Diego, San Diego, USA

Thomas Hornbein Papers (MSS669)
Ulrich Cameron Luft Papers (MSS475)
James Milledge Papers (MSS455)
L. G. C. E. Pugh Papers (MSS491)
John B. West Papers (MSS444)

University of Adelaide Rare Books & Special Collections, Adelaide, Australia.

W. V. Macfarlane Papers 1947–1985 (MS0006)

Wellcome Library, Archives & Manuscripts, London, UK

Explorers Cuttings Book (WF/M/GB/35)

PUBLISHED RESOURCES

Adam, R. I., and W. R. Stanmeyer. "Effects of Prolonged Antarctic Isolation on Oral and Intestinal Bacteria." *Oral Surgery, Oral Medicine, Oral Pathology* 13 (1960): 117–20.

Adler, Antony. "The Ship as Laboratory: Making Space for Field Science at Sea." *Journal of the History of Biology* (2013): 333–62.

Adolph, E. F., ed. *Physiology of Man in the Desert.* New York: Interscience Publishers, 1947.

Ambach, W., E. Ambach, W. Tributsch, R. Henn, and H. Unterdorfer. "Corpses Released from Glacier Ice: Glaciological and Forensic Aspects." *Journal of Wilderness Medicine* 3 (1992): 372–76.

Anderson, K. "Science and the Savage: The Linnean Society of New South Wales, 1874–1900." *Cultural Geographies* 5 (1998): 125–43.

Anderson, Warwick. "Climates of Opinion: Acclimatization in Nineteenth-Century France and England." *Victorian Studies* 35 (1992): 135–57.

———. *The Cultivation of Whiteness: Science, Health, and Racial Destiny in Australia*. Durham, NC: Duke University Press, 2006.

———. "Geography, Race and Nation: Remapping 'Tropical' Australia, 1890–1930." *Medical History Supplement* 44, S20 (2000): 146–59.

———. "Hybridity, Race, and Science: The Voyage of the *Zaca*, 1934–1935." *Isis* 103 (2012): 229–53.

———. "Immunities of Empire: Race, Disease, and the New Tropical Medicine, 1900–1920." *Bulletin of the History of Medicine* 70 (1996): 94–118.

———. "Where Every Prospect Pleases and Only Man Is Vile: Laboratory Medicine as Colonial Discourse." *Critical Inquiry* 18 (1992): 506–29.

Aragón, Santiago. "Le rayonnement international de la Société zoologique d'acclimatation: Participation de l'Espagne entre 1854 et 1861." *Revue d'histoire des sciences* 58 (2005): 169–206.

Arias-Stella, Javier. "Morphological Patterns: Mechanism of Pulmonary Arterial Hypertension." In *Life at High Altitudes* [Proceedings of the Special Session Held during the Fifth Meeting of the PAHO Advisory Committee on Medical Research, June 15, 1966], edited by A. Hurtado. Washington, DC: Pan American Health Organisation/World Health Organisation, 1966.

Arnold, David, ed. *Warm Climates and Western Medicine: The Emergence of Tropical Medicine, 1500–1900*. Amsterdam: Rodopi, 1996.

Aronova, Elena, Karen S. Baker, and Naomi Oreskes. "Big Science and Big Data in Biology: From the International Geophysical Year through the International Biological Program to the Long Term Ecological Research (LTER) Network, 1957–Present." *Historical Studies in the Natural Sciences* 40 (2010): 183–224.

Asahina, K. "Japanese Antarctic Expedition of 1911–12." In *Polar Human Biology: The Proceedings of the SCAR/IUPS/IUBS Symposium on Human Biology and Medicine in the Antarctic*, edited by O. G. Edholm and E. K. Eric Gunderson, 8–14. London: Heinemann Medical, 1973.

Aston, Felicity. "Women of the White Continent." *Geographical* 77 (2005): 26–30.

Aubin, David. "The Hotel That Became an Observatory: Mount Faulhorn as Singularity, Microcosm, and Macro-Tool." *Science in Context* 22 (2009): 365–86.

Auvinet, Gabriel, and Monique Briulet. "V. El Doctor Denis Jourdanet; Su Vida y Su Obra." *Gaceta Médica De México* 140 (2004): 426–29.

Barber, Noel. *The White Desert*. London: Hodder & Stoughton, 1958.

Barcroft, Joseph. "The Effect of Altitude on the Dissociation Curve of Blood." *Journal of Physiology* 42 (1911): 44–63.

———. "Recent Expedition to the Andes for the Study of the Physiology of High Altitudes (BAAS Section of Physiology)." *Lancet* 200 (1922): 685–86.

———. *The Respiratory Function of the Blood, Part I: Lessons from High Altitudes*. Cambridge: Cambridge University Press, 1925.

Barcroft, Joseph, C. A. Binger, A. V. Bock, J. H. Doggart, H. S. Forbes, G. Harrop, J. C. Meakins, and A. C. Redfield. "Observations upon the Effect of High Altitude on the Physiological Processes of the Human Body, Carried out in the Peruvian Andes, Chiefly at Cerro de Pasco." *Philosophical Transactions of the Royal Society of London, Series B* 211 (1923): 351–480.

Barrett, Frank A. "'Scurvy' Lind's Medical Geography." *Social Science & Medicine* 33 (1991): 347–53.

Battle, W. R. B. "An Experimental Concentrated Nut Food." *Polar Record* 7 (1954): 54.

Bauer, Paul. "Nanga Parbat, 1937." *Himalayan Journal* 10 (1938). https:///www.himalayanclub.org/jnl/10/9/nanga-parbat-1937/.

Beamish, Rob, and Ian Ritchie. "From Fixed Capacities to Performance-Enhancement: The Paradigm Shift in the Science of 'Training' and the Use of Performance-Enhancing Substances." *Sport in History* 25 (2005): 412–33.

Benson, Keith R., and Helen M. Rozwadowski, eds. *Extremes: Oceanography's Adventures at the Poles.* Sagamore Beach, MA: Science History Publications, 2007.

Berge, Anna, and Nello Pace. "Human Radiation Studies: Remembering the Early Years— Oral History of Physiologist Nello Pace Ph.D." United States Department of Energy— Office of Human Radiation Experiments—DOE/EH-0476, June 1995. http://www .iaea.org/inis/collection/NCLCollectionStore/_Public/27/059/27059281.pdf.

Berger, R. L. "Nazi Science—the Dachau Hypothermia Experiments." *New England Journal of Medicine* 322 (1990): 1435–40.

Bermúdez, José Luis. "Climbing on Kangchenjunga since 1955." *Alpine Journal* 101 (1996): 50–56.

Bert, Paul. *La pression barométrique. Recherches de physiologie expérimentale.* Paris: G. Masson, 1878.

"Biological and Medical Research Based on USS *Staten Island*, Antarctica, 1958–59." *Polar Record* 10 (1960): 146–48.

Blackburn, A. B. "Medical Research at Plateau Station." *Antarctic Journal of the United States* 3 (1968): 237–39.

Blakeney, T. S. "Maurice Wilson and Everest, 1934." *Alpine Journal* 70 (1965): 269–72.

Borden, Gail. *The Meat Biscuit: Invented, Patented, and Manufactured.* New York, 1853.

Boucher, Ellen. "Arctic Mysteries and Imperial Ambitions: The Hunt for Sir John Franklin and the Victorian Culture of Survival." *Journal of Modern History* 90 (2018): 40–75.

Bourdillon, T. D., "The Use of Oxygen Apparatus by Acclimatized Men." *Proceedings of the Royal Society of London, Series B* 143 (1954): 24–32.

Bourke, Joanna. "Pain Sensitivity: An Unnatural History from 1800 to 1965." *Journal of Medical Humanities* 35 (2014): 301–19.

Brotherhood, J. R. "The British Antarctic Environment with Special Reference to Energy Expenditure." *Forsvarsmedicine* 8 (1972): 111–14.

Brown, A. H. "Water Requirements of Man in the Desert." In *Physiology of Man in the Desert,* edited by E. F. Adolph, 115–35. New York: Interscience Publishers, 1947.

Brown, G. M., G. S. Bird, T. J. Boag, L. M. Boag, J. D. Delahaye, J. E. Green, J. D. Hatcher, and John Page. "The Circulation in Cold Acclimatization." *Circulation* 9 (1954): 813–22.

Budd, G. M. "Skin Temperature, Thermal Comfort, Sweating, Clothing and Activity of Men Sledging in Antarctica." *Journal of Physiology* 186 (1966): 201–15.

Budd, G. M., and N. Warhaft. "Cardiovascular and Metabolic Responses to Noradrenaline in Man, before and after Acclimatization to Cold in Antarctica." *Journal of Physiology* 186 (1966): 233–42.

Bull, Colin. "Behind the Scenes: Colin Bull Recalls His 10-Year Quest to Send Women Researchers to Antarctica." *Antarctic Sun,* November 13, 2009, https://antarcticsun.usap .gov/features/contentHandler.cfm?id=1955.

Burstyn, Harold L. "'Big Science' in Victorian Britain: The *Challenger* Expedition (1872– 76) and Its *Report* (1881–95)." In *Understanding the Oceans: A Century of Ocean Exploration,* edited by Margaret Deacong, Tony Rice, and Colin Summerhayes, 49–55. Boca Raton, FL: CRC Press, 2002.

Burton, Alan C., and Otto G. Edholm. *Man in a Cold Environment: Physiological and Pathological Effects of Exposure to Low Temperatures.* London: Edward Arnold, 1955.

Campbell, J. M. H., C. Gordon Douglas, and F. G. Hobson. "The Respiratory Exchange of Man during and after Muscular Exercise." *Philosophical Transactions of the Royal Society of London, Series B* 210 (1921): 1–47.

Capelotti, P. J. "Extreme Archaeological Sites and Their Tourism: A Conceptual Model from Historic American Polar Expeditions in Svalbard, Franz Josef Land and Northeast Greenland." *Polar Journal* 2 (2012): 236–55.

Cavell, Janice. "Going Native in the North: Reconsidering British Attitudes during the Franklin Search, 1848–1859." *Polar Record* 45 (2009): 25–35.

Cerretelli, P. "Limiting Factors to Oxygen Transport on Mount Everest." *Journal of Applied Physiology* 40 (1976): 658–67.

Cerretelli, P., A. Veicsteinas, and C. Marconi. "Anaerobic Metabolism at High Altitude: The Lactacid Mechanism." In *High Altitude Physiology and Medicine*, edited by W. Brendel and R. A. Zink, 94–102. New York: Springer-Verlag, 1982.

Chadarevian, Soraya de. "Human Population Studies and the World Health Organization." *Dynamis* 35 (2015): 359–88.

Chapman, Carleton B. "The Long Reach of Harvard's Fatigue Laboratory, 1926–1947." *Perspectives in Biology and Medicine* 34 (1990): 17–33.

Charnley, B. "Arguing over Adulteration: The Success of the Analytical Sanitary Commission." *Endeavour* 32 (2008): 129–33.

Cherry-Garrard, Apsley. *The Worst Journey in the World 1910–13.* London: Constable and Co., 1922.

Chiodi, Hugo. "Respiratory Adaptations to Chronic High Altitude Hypoxia." *Journal of Applied Physiology* 10 (1957): 81–87.

Clayton, Julie. *MRC National Institute for Medical Research: A Century of Science for Health.* London: MRC, 2014.

Clements, Philip W. *Science in an Extreme Environment: The 1963 American Mount Everest Expedition.* Pittsburgh: University of Pittsburgh Press, 2018.

Cloud, John, ed. "Earth Sciences in the Cold War." Special issue, *Social Studies of Science* 33, no. 5 (October 2003).

Coen, Deborah R. "The Storm Lab: Meteorology in the Austrian Alps." *Science in Context* 22 (2009): 463–86.

Collins, Harry. *Changing Order: Replication and Induction in Scientific Practice.* London: Sage Publications, 1985.

Collis, Christy. "Walking in Your Footsteps: 'Footsteps of the Explorers' Expeditions and the Contest for Australian Desert Space." In *New Spaces of Exploration: Geographies of Discovery in the Twentieth Century*, edited by S. Naylor and J. R. Ryan, 222–40. London: I. B. Tauris, 2009.

Colpitts, George. *Pemmican Empire: Food, Trade, and the Last Bison Hunts in the North American Plains, 1780–1882.* Cambridge: Cambridge University Press, 2015.

Cotes, J. E. "Ventilatory Capacity at Altitude and Its Relation to Mask Design." *Proceedings of the Royal Society of London, Series B* 143 (1954): 32–39.

Croft, Andrew. "West Greenland Sledge Dogs." *Polar Record* 2 (1937): 68–81.

Cueto, Marcos. "Andean Biology in Peru: Scientific Styles on the Periphery." *Isis* 80 (1989): 640–58.

———. "Laboratory Styles in Argentine Physiology." *Isis* 85 (1994): 228–46.

Cunningham, Andrew, and Perry Williams, eds. *The Laboratory Revolution in Medicine.* Cambridge: Cambridge University Press, 1992.

Curtin, Philip D. *Death by Migration: Europe's Encounter with the Tropical World in the Nineteenth Century.* Cambridge: Cambridge University Press, 1989.

Davenport, C. B., and Morris Steggerda. "Race Crossing in Jamaica." Washington, DC: Carnegie Institution, 1929.

Davis, A. G. "Seasonal Changes in Body Weight and Skinfold Thickness." *British Antarctic Survey Bulletin* 19 (1969): 75–81.

[Debenham, Frank?]. "The Eskimo Kayak." *Polar Record* 1 (1934): 52–62.

De Bont, Raf. "'Primitives' and Protected Areas: International Conservation and the 'Naturalization' of Indigenous People, ca. 1910–1975." *Journal of the History of Ideas* 76 (2015): 215–36.

———. *Stations in the Field: A History of Place-Based Animal Research, 1870–1930.* Chicago: University of Chicago Press, 2015.

DeC. W., R. "The Snow Huts of the Eskimo." *Bulletin of the American Geographical Society* 37 (1905): 674.

Deepak, Dass, and G. Bhaumik. "The Silver Hut Experiment." *Science Reporter* (November 2012), http://nopr.niscair.res.in/bitstream/123456789/15016/1/SR%2049%2811%29%2056-57.pdf.

Dent, Clinton T. "Can Mount Everest Be Ascended?" *Nineteenth Century* 32 (1892): 604–13.

Di Giulio, Camillo, and John B. West. "Angelo Mosso's Experiments at Very Low Barometric Pressures." *High Altitude Medicine & Biology* 14 (2013): 78–79.

Dill, D. B., and D. S. Evans. "Report Barometric Pressure!" *Journal of Applied Physiology* 29 (1970): 914–16.

Dimeo, Paul. *A History of Drug Use in Sport: 1876–1976: Beyond Good and Evil.* New York: Routledge/Taylor & Francis, 2007.

Dittert, René, Gabriel Chevalley, and Raymond Lambert. *Forerunners to Everest.* Translated by Malcolm Barnes. London: Hamilton & Co., 1956.

Douglas, Gordon, J. S. Haldane, Yandell Henderson, and Edward C. Schneider. "Physiological Observations Made on Pike's Peak, Colorado, with Special Reference to Adaptation to Low Barometric Pressures." *Philosophical Transactions of the Royal Society of London, Series B* 203 (1913): 185–318.

Driver, Felix. *Geography Militant: Cultures of Exploration and Empire.* Oxford: Oxford University Press, 2001.

Dufour, Marc. "Sur le mal de montagne." *Bulletin de la Société médicale de la Suisse Romande* 74 (1874): 72–79, 261–64.

Eagan, Charles J. "Resistance to Finger Cooling Related to Physical Fitness." *Nature* 200 (1963): 851–52.

Easty, D. L., A. Antonis, and I. Bersohn. "Adipose Fat Composition in Young Men in Antarctica." *British Antarctic Survey Bulletin* 13 (1967): 41–45.

Edholm, O. G. "Medical Research by the British Antarctic Survey." *Polar Record* 12 (1965): 575–82.

Edholm, O. G., and A. L. Bacharach, eds. *Exploration Medicine: Being a Practical Guide for Those Going on Expeditions.* Bristol: John Wright & Sons, 1965.

Edholm, O. G., and E. K. Eric Gunderson, eds. *Polar Human Biology: The Proceedings of the SCAR/IUPS/IUBS Symposium on Human Biology and Medicine in the Antarctic.* London: Heinemann Medical, 1973.

Edwards, H. T. "Lactic Acid in Rest and Work at High Altitude." *American Journal of Physiology* 116 (1936): 367–75.

Ellis, F. P., Helen M. Ferres, A. R. Lind, and P. S. B. Newling. "The Upper Limits of Tolerance of Environmental Stress." In *Physiological Responses to Hot Environments*, edited by R. K. Macpherson 158–79. London: HMSO, 1960.

Endersby, Jim. *Imperial Nature: Joseph Hooker and the Practices of Victorian Science.* Chicago: University of Chicago Press, 2008.

Eves, Richard. "Unsettling Settler Colonialism: Debates over Climate and Colonization in New Guinea, 1875–1914." *Ethnic and Racial Studies* 28 (2005): 304–30.

Farish, Matthew. "The Lab and the Land: Overcoming the Arctic in Cold War Alaska." *Isis* 104 (2013): 1–29.

Felsch, Philipp. *Laborlandschaften: Physiologische Alpenreisen Im 19. Jahrhundert.* Göttingen: Wallstein, 2007.

———. "Mountains of Sublimity, Mountains of Fatigue: Towards a History of Speechlessness in the Alps." *Science in Context* 22 (2009): 341–64.

FitzGerald, M. P. "The Changes in the Breathing and the Blood at Various High Altitudes." *Proceedings of the Royal Society of London, Series B* 88 (1913): 351–71.

FitzGerald, M. P., and G. Dreyer. *The Unreliability of the Neutral Red Method, as Generally Employed, for the Differentiation of* B. typhosus *and* B. coli. Reprinted from *Contributions from the University Laboratory for Medical Bacteriology [Copenhagen] to Celebrate the Inauguration of the State Serum Institute.* Copenhagen: O. C. Olsen & Co., 1902.

FitzGerald, M. P., and J. S. Haldane. "The Normal Alveolar Carbonic Acid Pressure in Man." *Journal of Physiology* 32 (1905): 486–94.

Fletcher, R. F. "Birmingham Medical Research Expeditionary Society 1977 Expedition: Signs and Symptoms." *Postgraduate Medical Journal* 55 (1979): 461–63.

Fogg, G. E. *A History of Antarctic Science.* Cambridge: Cambridge University Press, 1992.

Forsius, Henrik, Aldur W. Eriksson, and Johan Fellman. "The International Biological Program/Human Adaptability Studies among the Skolt Sami in Finland (1966–1970)." *International Journal of Circumpolar Health* 71 (2012): 1–5.

Fox, R. H., G. M. Budd, Patricia M. Woodward, A. J. Hackett, and A. L. Hendrie. "A Study of Temperature Regulation in New Guinea People." *Philosophical Transactions of the Royal Society of London, Series B* 268 (1974): 375–91.

Freeman, Milton M. R. "Adaptive Innovation among Recent Eskimo Immigrants in the Eastern Arctic Canada." *Polar Record* 14 (1969): 769–81.

Fuchs, Vivian. "The Commonwealth Trans-Antarctic Expedition." *Geographical Journal* 124 (1958): 439–50.

———. Foreword to *Man in the Antarctic: The Scientific Work of the International Biomedical Expedition to the Antarctic (IBEA)*, by Jean Rivolier. London: Taylor & Francis, 1988.

———. "Sledging Rations of the Falkland Islands Dependencies Survey, 1948–50." *Polar Record* 6 (1952): 508–11.

Gagge, A. P., A. C. Burton, and H. C. Bazett. "A Practical System of Units for the Description of the Heat Exchange of Man with His Environment." *Science* 94 (1941): 428–30.

Galton, F. *The Art of Travel.* 1st ed. London: John Murray, 1855.

———. *The Art of Travel.* 5th ed. London: John Murray, 1872.

Giaever, John. *The White Desert: The Official Account of the Norwegian-British-Swedish Antarctic Expedition.* London: Chatto & Windus, 1954.

Gieryn, Thomas F. "City as Truth-Spot." *Social Studies of Science* 36 (2006): 5–38.

Gilchrist, Paul. "The Politics of Totemic Sporting Heroes and the Conquest of Everest." *Anthropological Notebooks* 12 (2006): 35–52.

Gill, M. B., M. J. S. Milledge, L. G. C. E. Pugh, and J. B. West. "Alveolar Gas Composition at 21,000 to 25,700 Ft. (6400–7830 M)." *Journal of Physiology* 163 (1962): 373–77.

Gill, M. B., E. C. Poulton, A. Carpenter, M. M. Woodhead, and M. H. P. Gregory. "Falling Efficiency at Sorting Cards during Acclimatisation at 19,000 Ft." *Nature* 203 (1964): 436.

Gill, M. B., and L. G. C. E. Pugh. "Basal Metabolism and Respiration in Men Living at 5,800 m (19,000 Ft)." *Journal of Applied Physiology* 19 (1964): 949–54.

Gillespie, Richard. "Industrial Fatigue and the Discipline of Physiology." In *Physiology in the American Context, 1850–1940*, edited by Gerald L. Geison, 237–62. Bethesda, MD: American Physiological Society, 1987.

Giussani, D. A., L. Bennet, A. N. Sferruzzi-Perri, O. R. Vaughan, and A. L. Fowden. "Hypoxia, Fetal and Neonatal Physiology: 100 Years on from Sir Joseph Barcroft: Editorial." *Journal of Physiology* 594 (2016): 1105–11.

Glaser, E. M. "Acclimatization to Heat and Cold." *Journal of Physiology* 110 (1949): 330–37.

———. "Immersion and Survival in Cold Water." *Nature* 166 (1950): 1068.

Gleaves, John. "Manufactured Dope: How the 1984 US Olympic Cycling Team Rewrote the Rules on Drugs in Sports." *International Journal of the History of Sport* 32 (2015): 89–107.

Godin, G., and Roy J. Shephard. "Activity Patterns of the Canadian Eskimo." In *Polar Human Biology: The Proceedings of the SCAR/IUPS/IUBS Symposium on Human Biology and Medicine in the Antarctic*, edited by O. G. Edholm and E. K. Eric Gunderson, 193–215. London: Heinemann Medical, 1973.

Gooday, Graeme. "Placing or Replacing the Laboratory in the History of Science?" *Isis* 99 (2008): 783–95.

Greene, Raymond. "Observations on the Composition of Alveolar Air on Everest, 1933." *Journal of Physiology* 82 (1934): 481–85.

Greene, Shane. "Indigenous People Incorporated? Culture as Politics, Culture as Property in Pharmaceutical Bioprospecting." *Current Anthropology* 45 (2004): 211–37.

Grenfell Price, A. *White Settlers in the Tropics.* New York: American Geographical Society, 1939.

Guggenheim, M. "Laboratizing and De-Laboratizing the World: Changing Sociological Concepts for Places of Knowledge Production." *History of the Human Sciences* 25 (2012): 99–118.

Gugglberger, Martina. "Climbing Beyond the Summits: Social and Global Aspects of Women's Expeditions in the Himalayas." *International Journal of the History of Sport* 32 (2015): 597–613.

Guly, Henry Raymond. "Bacteriology during the Expeditions of the Heroic Age of Antarctic Exploration." *Polar Record* 49 (2013): 321–27.

———. "Frostbite and Other Cold Injuries in the Heroic Age of Antarctic Exploration." *Wilderness & Environmental Medicine* 23 (2012): 365–70.

———. "Human Biology Investigations during the Heroic Age of Antarctic Exploration (1897–1922)." *Polar Record* 50 (2014): 183–91.

———. "Medical Aspects of the Expeditions of the Heroic Age of Antarctic Exploration (1895–1922)." Ph.D. diss., University of Exeter, 2015.

———. "Medical Comforts during the Heroic Age of Antarctic Exploration." *Polar Record* 49 (2013): 110–17.

———. "Snow Blindness and Other Eye Problems during the Heroic Age of Antarctic Exploration." *Wilderness & Environmental Medicine* 23 (2012): 77–82.

———. "Surgery and Anaesthesia during the Heroic Age of Antarctic Exploration (1895–1922)." *British Medical Journal* 347 (December 17, 2013): f7242.

Gunderson, E. K. Eric. "Psychological Studies in Antarctica." In *Human Adaptability to Antarctic Conditions*, ed. E. K. Eric Gunderson, 225–31. Antarctic Research Series, vol. 22. Washington, DC: American Geophysical Union, 1974.

Gunga, Hans-Christian. *Nathan Zuntz: His Life and Work in the Fields of High Altitude Physiology and Aviation Medicine.* New York: Springer Verlag, 2008.

Hackett, Peter H., John T. Reeves, Charlotte D. Reeves, Robert F. Grover, and Drummond Rennie. "Control of Breathing in Sherpas at Low and High Altitude." *Journal of Applied Physiology* 49 (1980): 374–79.

Hagen, Joel B. "Bergmann's Rule, Adaptation, and Thermoregulation in Arctic Animals: Conflicting Perspectives from Physiology, Evolutionary Biology, and Physical Anthropology after World War II." *Journal of the History of Biology* 50 (2017): 235–65.

Haldane, J. B., J. Barcroft, Leonard Hill, Professor Boycott, T. G. Longstaff, Malcolm L. Hepburn, and Dr. Price-Jones. "Physiological Difficulties in the Ascent of Mount Everest: Discussion." *Geographical Journal* 65 (1925): 16–23.

Haldane, J. S. "Acclimatisation to High Altitudes." *Physiological Reviews* VII, 3 (1927): 363–84.

Haldane, J. S., A. M. Kellas, and E. L. Kennaway. "Experiments on Acclimatisation to Reduced Atmospheric Pressure." *Journal of Physiology* 53 (1919): 181–206.

Halsey, L. G., and M. A. Stroud. "100 Years since Scott Reached the Pole: A Century of Learning about the Physiological Demands of Antarctica." *Physiological Reviews* 92 (2012): 521–36.

Hamlin, Jacob. *Oceanographers and the Cold War: Disciplines of Marine Science*. Seattle: University of Washington Press, 2005.

Hampton, I. F. G. "Local Acclimatisation of the Hands to Prolonged Cold Exposure in the Antarctic." *British Antarctic Survey Bulletin* 19 (1969): 9–56.

Handford, Jenny Mai. "Dog Sledging in the Eighteenth Century: North America and Siberia." *Polar Record* 34 (1998): 237–48.

Harrison, Mark. *Climates & Constitutions: Health, Race, Environment and British Imperialism in India, 1600–1850*. Oxford: Oxford University Press, 1999.

———. "'The Tender Frame of Man': Disease, Climate and Racial Difference in India and the West Indies, 1760–1860." *Bulletin of the History of Medicine* 70 (1996): 68–93.

Heggie, Vanessa. "Experimental Physiology, Everest and Oxygen: From the Ghastly Kitchens to the Gasping Lung." *British Journal for the History of Science* 46 (2013): 123–47.

———. *A History of British Sports Medicine*. Manchester: Manchester University Press, 2011.

———. "Introduction: Special Section—Harvard Fatigue Laboratory." *Journal of the History of Biology* 48 (2015): 361–64.

———. "'Only the British Appear to Be Making a Fuss': The Science of Success and the Myth of Amateurism at the Mexico Olympiad, 1968." *Sport in History* 28 (2008): 213–35.

———. "Rationalised and Rationed: Food and Health." In *Cultural History of Medicine: Age of Empire, 1800–1920*, edited by J. Reinarz. London: Bloomsbury, forthcoming.

———. "Why Isn't Exploration a Science?" *Isis* 105 (2014): 318–34.

Hellon, R. F., R. M. Jones, R. K. Macpherson, and J. S. Weiner. "Natural and Artificial Acclimatisation to Hot Environments." *Journal of Physiology* 132 (1956): 559–76.

Hill, A. V. *Muscular Movement in Man: The Factors Governing Speed and Recovery from Fatigue*. New York: McGraw-Hill, 1927.

Hillary, Edmund, and Desmond Doig. *High in the Thin Cold Air*. London: Hodder & Stoughton, 1963.

Hingston, R. W. G. "Physiological Difficulties in the Ascent of Mount Everest." *Geographical Journal* 65 (1925): 4–16.

Hoberman, John M. *Mortal Engines: The Science of Performance and the Dehumanization of Sport*. Caldwell, NJ: Blackburn Press, 2001.

Hodacs, Hanna. "In the Field: Exploring Nature with Carolus Linnaeus." *Endeavour* 34 (2010): 45–49.

———. "Linnaeans Outdoors: The Transformative Role of Studying Nature 'on the Move' and Outside." *British Journal for the History of Science* 44 (2011): 183–209.

Hoebusch, Harald. "Ascent into Darkness: German Himalaya Expeditions and the National Socialist Quest for High-Altitude Flight." *International Journal of the History of Sport* 24 (2007): 520–40.

Hohwu Christensen, E. "Respiratory Control in Acute and Prolonged Hypoxia." *Proceedings of the Royal Society of London, Series B* 143 (1954): 8–12.

Hong, Suk Ki, and Hermann Rahn. "The Diving Women of Korea and Japan." In *Human Physiology and the Environment in Health and Disease: Readings from Scientific American*, edited by Arthur J. Vander, chap. 9. San Francisco: W. H. Freeman, 1976.

Horscroft, James A., Aleksandra O. Kotwica, Verena Laner, James A. West, Philip J. Hennis, Denny Z. H. Levett, David J. Howard et al. "Metabolic Basis to Sherpa Altitude Adaptation." *Proceedings of the National Academy of Sciences* 114 (2017): 6382–87.

Horvath, Steven M., and Elizabeth C. Horvath. *The Harvard Fatigue Laboratory: Its History and Contributions*. International Research Monograph Series in Physical Education. Englewood Cliffs, NJ: Prentice-Hall, 1973.

Houston, Charles S. "Introductory Address: Lessons to Be Learned from High Altitude." *Postgraduate Medical Journal* 55 (July 1979): 447–53.

———. "Operation Everest: A Study of Acclimatization to Anoxia." *US Naval Medical Bulletin* 46 (1946): 1783–92.

Houston, Charles S., and R. L. Riley. "Respiratory and Circulatory Changes during Acclimatisation to High Altitude." *American Journal of Physiology* 149 (1947): 565–88.

Hoyland, G. "Testing Mallory's Clothes on Everest." *Alpine Journal* 112 (2007): 243–46.

Hunt, John. *The Ascent of Everest*. London: Hodder & Stoughton, 1953.

Hurtado, A. "Natural Acclimatization to High Altitudes: Review of Concepts." In *Life at High Altitudes* [Proceedings of the Special Session Held during the Fifth Meeting of the PAHO Advisory Committee on Medical Research, June 15, 1966], edited by A. Hurtado. Washington, DC: Pan American Health Organisation/World Health Organisation, 1966.

"Igloos in the Alps." *Polar Record* 3 (1942): 512–16.

Itoh, Frederick A. "Physiology of Circumpolar People." In *The Human Biology of Circumpolar Populations*, edited by Frederick A. Milan, 285–303. Cambridge: Cambridge University Press, 1980.

Jackson, Frederic. "The Heart at High Altitude." *British Heart Journal* 30 (1968): 291–94.

Jackson, Frederic, and Hywel Davies. "The Electrocardiogram of the Mountaineer at High Altitude." *British Heart Journal* 22 (1960): 671–85.

Jackson, Mark. *The Age of Stress: Science and the Search for Stability*. Oxford: Oxford University Press, 2013.

Jackson, Monica, and Elizabeth Stark. *Tents in the Clouds: The First Women's Himalayan Expedition*. London: Travel Book Club, 1957.

Jacobson, A. *Operation Paperclip: The Secret Intelligence Program That Brought Nazi Scientists to America*. Boston: Little, Brown, 2014.

Jennings, Eric. T. *Curing the Colonizers: Hydrotherapy, Climatology, and French Colonial Spas*. Durham, NC: Duke University Press, 2006.

Johnson, Ryan. "European Cloth and 'Tropical' Skin: Clothing Material and British Ideas of Health and Hygiene in Tropical Climates." *Bulletin of the History of Medicine* 83 (2009): 530–60.

Jones, J. H. Emlyn. "Ama Dablam, 1959." *Alpine Journal* 65 (1960): 1–10.

Jones, Max. "From 'Noble Example' to 'Potty Pioneer': Rethinking Scott of the Antarctic, c. 1945–2011." *Polar Journal* 1 (2011): 191–206.

———. *The Last Great Quest: Captain Scott's Antarctic Sacrifice.* Oxford: Oxford University Press, 2004.

Joseph, Michael. "Military Officers, Tropical Medicine, and Racial Thought in the Formation of the West India Regiments, 1793–1802." *Journal of the History of Medicine and Allied Sciences* 72 (2016): 142–65.

Kaaij, J. M. van der, D. S. Martin, M. G. Mythen, and M. P. Grocott. "Research on Mount Everest: Exploring Adaptation to Hypoxia to Benefit the Critically Ill Patient." *Netherlands Journal of Critical Care* 15 (2011): 240–48.

Kaiser, David. *How the Hippies Saved Physics: Science, Counterculture and the Quantum Revival.* London: Norton, 2011.

Karafantis, Layne. "Sealab II and Skylab: Psychological Fieldwork in Extreme Spaces." *Historical Studies in the Natural Sciences* 43 (2013): 551–88.

Keatinge, W. R. "The Effect of Repeated Daily Exposure to Cold and of Improved Physical Fitness on the Metabolic and Vascular Response to Cold Air." *Journal of Physiology* 157 (1961): 209–20.

Keighren, Innes M. "A Scot of the Antarctic: The Reception and Commemoration of William Speirs Bruce." Master's thesis, University of Edinburgh, 2003.

Kellas, A. M. "A Consideration of the Possibility of Ascending the Loftier Himalaya." *Geographical Journal* 49 (1917): 26–46.

———. "Dr. Kellas' Expedition to Kamet." *Geographical Journal* 57 (1921): 124–30.

Kellogg, R. H. "'La Pression Barométrique': Paul Bert's Hypoxia Theory and Its Critics." *Respiration Physiology* 34 (1978): 1–28.

Kennedy, Dane. "British Exploration in the Nineteenth Century: A Historiographical Survey." *History Compass* 5 (2007): 1879–1900.

———. *The Magic Mountains: Hill Stations and the British Raj.* Berkeley: University of California Press, 1996.

Kenney, W. Larry, David W. DeGroot, and Lacy Alexander Holowatz. "Extremes of Human Heat Tolerance: Life at the Precipice of Thermoregulatory Failure." *Journal of Thermal Biology* 29 (2004): 479–85.

Kenny, J. T. "Claiming the High Ground: Theories of Imperial Authority and the British Hill Stations in India." *Political Geography* 16 (1997): 655–73.

———. "Climate, Race, and Imperial Authority: The Symbolic Landscape of the British Hill Station in India." *Annals of the Association of American Geographers* 85 (1995): 694–714.

Kestner, O. "Klimatologische Studien. L Der wirksame Anteil des Hoheriklimas." *Zeitschrift für Biologie* 73 (1921): 1–6.

Keys, Ancel. "The Physiology of Life at High Altitudes." *Scientific Monthly* 43 (1936): 289–312.

Keys, Jennifer, and Henry Guly. "The Medical History of South Georgia." *Polar Record* 45 (2009): 269–73.

Koerner, Lisbet. "Purposes of Linnean Travel: A Preliminary Research Report." In *Visions of Empire: Voyages, Botany and the Representation of Nature*, edited by David Miller and Peter Reill, 117–52. Cambridge: Cambridge University Press, 2011.

Kohler, Robert E. "Lab History: Reflections." *Isis* 99 (2008): 761–68.

———. "Labscapes: Naturalizing the Lab." *History of Science* 40 (2008): 473–501.

———. *Landscapes and Labscapes: Exploring the Lab-Field Border in Biology.* Chicago: University of Chicago Press, 2002.

———. "Place and Practice in Field Biology." *History of Science* 40 (2002): 189–210.

———. "Practice and Place in Twentieth-Century Field Biology: A Comment." *Journal of the History of Biology* 45 (2012): 579–86.

Kreuzer, F., and Z. Turek. "Influence of the Position of the Oxygen Dissociation Curve on the Oxygen Supply to Tissues." In *High Altitude Physiology and Medicine*, edited by W. Brendel and R. A. Zink, 66–72. New York: Springer-Verlag, 1982.

Kroll, Gary. *America's Ocean Wilderness: A Cultural History of Twentieth-Century Exploration.* Lawrence: University of Kansas Press, 2008.

Kuklick, Henrika. "Personal Equations: Reflections on the History of Fieldwork, with Special Reference to Sociocultural Anthropology." *Isis* 102 (2011): 1–33.

Lahiri, S., and J. S. Milledge. "Acid-Base in Sherpa Altitude Residents and Lowlanders at 4880 m." *Respiration Physiology* 2 (1967): 323–34.

———. "Sherpa Physiology." *Nature* 207 (1965): 610–12.

Landy, David. "Pibloktoq (Hysteria) and Inuit Nutrition: Possible Implication of Hypervitaminosis A." *Social Science & Medicine* 21 (1985): 173–85.

Latham, D. V., and C. Gillman. "Kilimanjaro and Some Observations on the Physiology of High Altitudes in the Tropics." *Geographical Journal* 68 (1926): 492–505.

Latour, Bruno. *The Pasteurisation of France.* Translated by Alan Sheridan and John Law. Cambridge, MA: Harvard University Press, 1988.

Le Gars, Stephane, and David Aubin. "The Elusive Placelessness of the Mont-Blanc Observatory (1893–1909): The Social Underpinnings of High-Altitude Observation." *Science in Context* 22 (2009): 509–31.

Lester, James T. *Behavioral Research during the 1963 American Mount Everest Expedition.* Final Report September 1964. www.dtic.mil/dtic/tr/fulltext/u2/607336.pdf.

Lever, Christopher. *They Dined on Eland: The Story of the Acclimatisation Societies.* London: Quiller Press, 1999.

Lewis, H. E., and J. P. Masterton. "British North Greenland Expedition 1952–54: Medical and Physiological Aspects [Part 1]." *Lancet* 266 (September 3, 1955): 494–500.

———. "British North Greenland Expedition 1952–54: Medical and Physiological Aspects [Part 2]." *Lancet* 266 (1955): 549–56.

Livingstone, David N. "Changing Climate, Human Evolution, and the Revival of Environmental Determinism." *Bulletin of the History of Medicine* 86 (2012): 564–95.

———. "The Moral Discourse of Climate: Historical Considerations on Race, Place and Virtue." *Journal of Historical Geography* 17 (1991): 413–34.

Lloyd, R. M. "Ketonuria in the Antarctic: A Detailed Study." *British Antarctic Survey Bulletin* 20 (1969): 59–68.

Lobban, Mary C. "Cambridge Spitsbergen Physiological Expedition, 1953." *Polar Record* 48 (1954): 151–61.

Lobenhoffer, H. P., R. A. Zink, and W. Brendel. "High Altitude Pulmonary Edema: Analysis of 166 Cases." In *High Altitude Physiology and Medicine*, W. Brendel and R. A. Zink, 219–31. New York: Springer-Verlag, 1982.

Longstaff, T. G. "A Mountaineering Expedition to the Himalaya of Garhwal." *Geographical Journal* 31 (1908): 361–88.

———. *Mountain Sickness and Its Probable Causes.* London: Spottiswoode, 1906.

Lossio, Jorge. "Life at High Altitudes: Medical Historical Debates (Andean Region, 1890–1960)." Ph.D. diss., University of Manchester, 2006.

Lowe, George. *From Everest to the South Pole.* New York: St. Martin's Press, 1961. [Originally published in 1959 under the title *Because It Is There*.]

Malavielle, Louis. "Vacances en igloo sur le Mont-Blanc." *La Montaigne* vii (May 1939): 141–51.

Markham, Clements R. *The Lands of Silence.* Cambridge: Cambridge University Press, 1921.

Martin, Emily. "The Egg and the Sperm: How Science Has Constructed a Romance Based on Stereotypical Male-Female Roles." *Signs* 16 (1991): 485–501.

Matthews, D. S., B. P. Tribedi, A. R. Roy, R. Chatterjee, and A. Ghosal. "Some Effects of High-Altitude Climbing; Investigations Made on Climbers of the British Kangchenjunga Reconnaissance Expedition, 1954." *British Medical Journal* 1 (March 26, 1955): 768–69.

Mawson, Douglas. *Home of the Blizzard.* London: St. Martin's Press, 1999.

McCormick, Iain A., Antony J. W. Taylor, Jean Rivolier, and Genevieve Cazes. "A Psychometric Study of Stress and Coping during the International Biomedical Expedition to the Antarctic (IBEA)." *Journal of Human Stress* 11 (1985): 150–56.

Merson, John. "Bio-Prospecting or Bio-Piracy: Intellectual Property Rights and Biodiversity in a Colonial and Postcolonial Context." *Osiris* 15 (2000): 282–96.

Messmer, K. "Oxygen Transport Capacity." In *High Altitude Physiology and Medicine*, edited by W. Brendel and R. A. Zink, 21–27. New York: Springer-Verlag, 1982.

Milledge, J. S. "Electrocardiographic Changes at High Altitude." *British Heart Journal* 25 (1963): 291–98.

———. "The Great Oxygen Secretion Controversy." *Lancet* 326 (1985): 1408–11.

———. "The Silver Hut Expedition, 1960–1961." *High Altitude Medicine & Biology* 11 (2010): 93–101.

Mitchell, H. H., and Marjorie Edman. *Nutrition and Resistance to Climatic Stress with Particular Reference to Man.* Report: Quartermaster Food and Container Institute for the Armed Forces, Research and Development Branch, November 1949. www.dtic.mil /dtic/tr/fulltext/u2/a581922.pdf.

Monge Medrano, Carlos. *Acclimatization in the Andes.* Baltimore: Johns Hopkins University Press, 1948.

Moreno, Jonathan D. *Undue Risk: Secret State Experiments on Humans.* New York: W. H. Freeman, 1999.

Morin, Nea. *A Woman's Reach: Mountaineering Memoirs.* London: Eyre & Spottiswoode, 1968.

Morshead, H. T. "Report on the Expedition to Kamet, 1920." *Geographical Journal* 57 (1921): 213–19.

Morus, Iwan Rhys. "Invisible Technicians, Instrument Makers and Artisans." In *A Companion to the History of Science*, edited by B. Lightman, 97–110. London: Wiley Blackwell, 2016.

Mosso, Angelo. *Der Mensch auf den Hochalpen.* Leipzig: Verlag von Veit & Comp., 1899.

———. *Life of Man on the High Alps.* Translated by E. Lough Kiesow. London: T. Fisher Unwin, 1898.

Mulgrew, Peter. *No Place for Men.* Auckland: Longman Paul, 1981.

Murray, Carl. "The Use and Abuse of Dogs on Scott's and Amundsen's South Pole Expeditions." *Polar Record* 44 (2008): 303–10.

Naylor, S., and J. Ryan, eds. *New Spaces of Exploration: Geographies of Discovery in the Twentieth Century.* London: I. B. Tauris, 2009.

Neill, Deborah. "Finding the 'Ideal Diet': Nutrition, Culture, and Dietary Practices in France and French Equatorial Africa, c. 1890s to 1920s." *Food and Foodways* 17 (2009): 1–28.

Newburgh, L. H., ed. *The Physiology of Heat Regulation and the Science of Clothing Prepared at the Request of the Division of Medical Sciences, National Research Council.* Philadelphia: W. B. Saunders, 1949.

Nicely, Polly G., and Judith K. Childers. "Mt. Everest Reveals Its Secrets to Medicine and Science: A Report on the 1981 American Medical Research Expedition to Everest." *Journal of the Indiana State Medical Association* 75 (1982): 704–8.

Nielsen, Hanne E. F., "Hoofprints in Antarctica: Byrd, Media, and the Golden Guernseys." *Polar Journal* 6 (2016): 342–57.

Norman, J. N. "Cold Exposure and Patterns of Activity at a Polar Station." *British Antarctic Survey Bulletin* 6 (1965): 1–13.

Noyce, Wilfred. *South Col: One Man's Adventure on the Ascent of Everest 1953.* London: Reprint Society, 1955.

"Obituary: Alexander Mitchell Kellas, D.Sc. (Lond.), Ph.D. (Heidelberg)." *Geographical Journal* 58 (1921): 73–75.

Oelz, Oswald. "How to Stay Healthy While Climbing Mount Everest." In *High Altitude Physiology and Medicine*, edited by W. Brendel and R. A. Zink, 298–300. New York: Springer-Verlag, 1982.

Ophir, A. "The Place of Knowledge: A Methodological Survey." *Science in Context* 4 (1991): 3–21.

Oreskes, Naomi. "Objectivity or Heroism? On the Invisibility of Women in Science." *Osiris* 11 (1996): 87–113.

Ortner, Sherry B. *Life and Death on Mt. Everest: Sherpas and Himalayan Mountaineering.* Princeton, NJ: Princeton University Press, 1999.

———. "Thick Resistance: Death and the Cultural Construction of Agency in Himalayan Mountaineering." *Representations* 59 (1997): 135–62.

Osborne, Michael. "Acclimatizing the World: A History of the Paradigmatic Colonial Science." *Osiris* 15 (2000): 135–51.

———. *The Emergence of Tropical Medicine in France.* Chicago: University of Chicago Press, 2014.

Osseo-Asare, Abena Dove. "Bioprospecting and Resistance: Transforming Poisoned Arrows into Strophantin Pills in Colonial Gold Coast, 1885–1922." *Social History of Medicine* 21 (2008): 269–90.

———. *Bitter Roots: The Search for Healing Plants in Africa.* Chicago: University of Chicago Press, 2014.

Outram, Dorinda. "On Being Perseus: New Knowledge, Dislocation, and Enlightenment Exploration." In *Geography and Enlightenment*, edited by David N. Livingstone and Charles W. J. Withers, 281–94. Chicago: University of Chicago Press, 1997.

Pace, Nello, L. Bruce Meyer, and Burton E. Vaughan. "Erythrolysis on Return of Altitude Acclimatized Individuals to Sea-Level." *Journal of Applied Physiology* 9 (1956): 141–44.

Palinkas, L. A., and D. Browner. "Stress, Coping and Depression in US Antarctic Program Personnel." *Antarctic Journal of the United States* 26 (1991): 240–41.

Parsons, Mike, and Mary B. Rose. *Invisible on Everest: Innovation and the Gear Makers.* London: Old City Publishing, 2002.

Peary, Robert E. *Secrets of Polar Travel.* New York: Century Co., 1917. Reprint, Elibrion Classics, 2007.

"Physiology of Channel Swimmers." *British Medical Journal* 2 (September 3, 1960): 725.

Pickman, Sarah. "Dress, Image, and Cultural Encounter in the Heroic Age of Polar Exploration." In *Expedition: Fashion from the Extreme*, 31–56. New York: Thames & Hudson, 2017.

Pickstone, John V. "Museological Science? The Place of the Analytical/Comparative in Nineteenth-Century Science, Technology and Medicine," *History of Science* 31 (1994): 111–38.

———. *Ways of Knowing: A New History of Science, Technology, and Medicine.* Chicago: University of Chicago Press, 2001.

Pohl-Valero, Stefan. "¿Agresiones de la altura y degeneración fisiológica? La biografía del 'clima' como objeto de investigación científica en Colombia durante el siglo XIX e inicios del XX." In *Historias alternativas de la fisiología en América Latina*, número especial, *Revista Ciencias de la Salud* 13 (2015): 65–83.

———. "'La raza entra por la boca': Energy, Diet, and Eugenics in Colombia, 1890–1940." *Hispanic American Historical Review* 94 (2014): 455–86.

"Polar Medicine." *Lancet* 274 (November 7, 1959): 786–87.

Pols, Hans. "Notes from Batavia, the Europeans' Graveyard: The Nineteenth-Century Debate on Acclimatization in the Dutch East Indies." *Journal of the History of Medicine and Allied Sciences* 67 (2012): 120–48.

Priestley, Raymond. "Twentieth-Century Man against Antarctica." *Nature* 178 (September 1, 1956): 463–70.

Pugh, L. G. C. E. "Accidental Hypothermia in Walkers, Climbers, and Campers: Report to the Medical Commission on Accident Prevention." *British Medical Journal* 1 (January 15, 1966).

———. "Carbon Monoxide Content of the Blood and Other Observations on Weddell Seals." *Nature* 183 (1959): 74–76.

———. "Carbon Monoxide Hazard in Antarctica." *British Medical Journal* 1 (January 24, 1959): 192–96.

———. "Clothing Insulation and Accidental Hypothermia in Youth." *Nature* 209 (1966): 1281–86.

———. "Cold Stress and Muscular Exercise, with Special Reference to Accidental Hypothermia." *British Medical Journal* 2 (May 6, 1967): 333–37.

———. "The Effects of Oxygen on Acclimatized Men at High Altitude." *Proceedings of the Royal Society of London, Series B* 143 (1954): 14–17.

———. "Haemoglobin Levels in the British Himalayan Expeditions to Cho Oyu in 1952 and Everest in 1953." *Journal of Physiology* 126-Supp. (1954): 38–39.

———. "Muscular Exercise on Mount Everest." In *High Altitude Physiology: Benchmark Papers in Human Physiology*, edited by John West, 77–109. Stroudsburg, PA: Hutchinson Ross, 1981.

———. "Muscular Exercise on Mount Everest." *Journal of Physiology* 141 (1958): 233–61.

———. "Physiological and Medical Aspects of the Himalayan Scientific and Mountaineering Expedition." *British Medical Journal* 2 (September 8, 1962): 621–27.

———. "Resting Ventilation and Alveolar Air on Mount Everest." *Journal of Physiology* 135 (1957): 590–610.

———. "Science in the Himalaya." *Nature* 191 (1961): 429–30.

———. "Tolerance to Extreme Cold at Altitude in a Nepalese Pilgrim." *Journal of Applied Physiology* 18 (1963): 1234–38.

Pugh, L. G. C. E., and O. G. Edholm. "The Physiology of Channel Swimmers." *Lancet* 269 (1955): 761–68.

Rabinbach, Anson. *The Human Motor: Energy, Fatigue and the Origins of Modernity.* Berkeley: University of California Press, 1992.

Rack, Ursula. "Felix König and the European Science Community across Enemy Lines during the First World War." *Polar Journal* 4 (2014): 88–104.

Radin, Joanna. "Latent Life: Concepts and Practices of Human Tissue Preservation in the International Biological Program." *Social Studies of Science* 43 (2013): 484–508.

———. *Life on Ice: A History of New Uses for Cold Blood.* Chicago: University of Chicago Press, 2017.

Rahn, Hermann. "Lessons from Breath Holding." In *The Regulation of Human Respiration: The Proceedings of the J. S. Haldane Centenary Symposium Held in the University Laboratory of Physiology, Oxford*, edited by D. J Cunningham and B. B. Lloyd. Oxford: Blackwell Scientific, 1963.

Raj, Kapil. *Relocating Modern Science: Circulation and the Construction of Knowledge in South Asia and Europe, 1650–1900*. London: Routledge, 2007.

———. "When Human Travellers Become Instruments." In *Instruments, Travel and Science*, edited by Marie Noëlle Bourguet, Christian Licoppe, and H. Otto Sibum, 156–88. London: Routledge, 2002.

Ray, Arthur J. "The Northern Great Plains: Pantry of the Northwestern Fur Trade, 1774–1885." *Prairie Forum* 9 (1984): 263–80.

Reeves, John T., B. M. Groves, J. R. Sutton, P. D. Wagner, A. Cymerman, M. K. Malconian, P. B. Rock, P. M. Young, and C. S. Houston. "Operation Everest II: Preservation of Cardiac Function at Extreme Altitude." *Journal of Applied Physiology* 63 (1987): 531–39.

Reeves, Nicky. "'To Demonstrate the Exactness of the Instrument': Mountainside Trials of Precision in Scotland, 1774." *Science in Context* 22 (2009): 323–40.

Richalet, Jean-Paul. "Operation Everest III: COMEX '97." *High Altitude Medicine & Biology* 11 (2010): 121–32.

Riffenburgh, Beau, ed. *Encyclopaedia of the Antarctic*. Vol. 1. London: Routledge 2007.

———. *The Myth of the Explorer: The Press, Sensationalism and Geographical Discovery*. London: Belhaven Press, 1993.

Rivolier, Jean. *Man in the Antarctic: The Scientific Work of the International Biomedical Expedition to the Antarctic (IBEA)*. London: Taylor & Francis, 1988.

———. "Physiological and Psychological Studies Conducted by Continental European and Japanese Expeditions." In *Human Adaptability to Antarctic Conditions*, edited by E. K. Eric Gunderson, 55–70. Antarctic Research Series, vol. 22. Washington, DC: American Geophysical Union, 1974.

Roberts, Peder. *The European Antarctic: Science and Strategy in Scandinavia and the British Empire*. New York: Palgrave Macmillan, 2011.

———. "Heroes for the Past and Present: A Century of Remembering Amundsen and Scott." In "Beyond the Limits of Latitude: Reappraising the Race to the South Pole," special issue, *Endeavour* 35 (2011): 142–50.

Roberts, Peder, and Dolly Jørgensen. "Animals as Instruments of Norwegian Imperial Authority in the Interwar Arctic." *Journal for the History of Environment and Society* 1 (2016): 65–87.

Robinson, Douglas H. *The Dangerous Sky: A History of Aviation Medicine*. Henley-on-Thames: Foulis, 1973.

Robinson, Michael F. *The Coldest Crucible: Arctic Exploration and American Culture*. Chicago: University of Chicago Press, 2006.

Roche, Clare. "Women Climbers 1850–1900: A Challenge to Male Hegemony?" *Sport in History* 3 (2013): 236–59.

Rodahl, Kåre. *Between Two Worlds: A Doctor's Log-Book of Life amongst the Alaskan Eskimos*. 2nd ed. London: Heinemann, 1964.

Rodahl, Kåre, and T. Moore. "The Vitamin A Content and Toxicity of Bear and Seal Liver." *Biochemical Journal* 37 (1943): 166–68.

Rodríguez de Romo, Ana Cecilia, and José Rogelio Pérez Padillia. "The Mexican Response to High Altitudes in the 1890s: The Case of a Physician and His 'Magic Mountain.'" *Medical History* 42 (2003): 493–516.

Rodway, George W. "Historical Vignette: George Ingle Finch and the Mount Everest Expedition of 1922: Breaching the 8000-m Barrier." *High Altitude Medicine & Biology* 8 (2007): 68–76.

———. "Mountain Clothing and Thermoregulation: A Look Back." *Wilderness & Environmental Medicine* 23 (2012): 91–94.

——. "Prelude to Everest: Alexander M. Kellas and the 1920 High Altitude Scientific Expedition to Kamet." *High Altitude Medicine & Biology* 5 (2004): 364–79.

——. "Ulrich C. Luft and Physiology on Nanga Parbat: The Winds of War." *High Altitude Medicine & Biology* 10 (2009): 89–96.

Rogers, A. F., and R. J. Sutherland. *Antarctic Climate, Clothing and Acclimatization: Final Scientific Report.* Bristol: Bristol University Department of Physiology, 1971.

Rose, Mary B., and Mike Parsons. *Mallory Myths and Mysteries: The Mallory Clothing Replica Project.* Keswick, Cumbria: Mountain Heritage Trust, 2006.

Ross, Helen E. "Sleep and Wakefulness in the Arctic under an Irregular Regime." In *Biometeorology. Proceedings of the Second International Bioclimatological Conference (1960)*, edited by S. W. Tromp, 389–94. Oxford: Pergamon Press, 1962.

Rostène, William. "Paul Bert: homme de science, homme politique." *Journal de la Société de Biologie* 200 (2006): 245–50.

Roth, Karl Heinz. "Flying Bodies—Enforcing States: German Aviation Medical Research from 1925 to 1975 and the Deutsche Forschungsgemeinschaft." In *Man, Medicine, and the State: The Human Body as an Object of Government Sponsored Medical Research in the 20th Century*, edited by Wolfgang U. Eckart, 108–32. Stuttgart: Franz Steiner Verlag Wiesbaden GmbH, 2006.

Rowley, Graham. "Snow-House Building." *Polar Record* 2 (1938): 109–16.

Roxburgh, H. L. "Oxygen Equipment for Climbing Mount Everest." *Geographical Journal* 109 (1947): 207–16.

Rozwadowski, Helen M. *Fathoming the Ocean: The Discovery and Exploration of the Deep Sea.* Cambridge, MA: Harvard University Press, 2005.

Ruttledge, Hugh. *Everest: The Unfinished Adventure.* London: Hodder & Stoughton, 1937.

——. "The Mount Everest Expedition, 1933." *Geographical Journal* 83 (1934): 1–9.

Sánchez Arteaga, Juanma. "Biological Discourses on Human Races and Scientific Racism in Brazil (1832–1911)." *Journal of the History of Biology* 50 (2017): 267–314.

Sarnquist, F. H. "Physicians on Mount Everest." *Western Journal of Medicine* 139 (1983): 480–85.

Sarnquist, F. H., R. Schoene, and P. Hackett. "Exercise Tolerance and Cerebral Function after Acute Hemodilution of Polycythemic Mountain Climbers." *Physiologist* 25 (1982): 327.

Savourey, G., A. Guinet, Y. Besnard, N. Garcia, A. Hanniquet, and J. Bittel. "Are the Laboratory and Field Conditions Observations of Acute Mountain Sickness Related?" *Aviation, Space, and Environmental Medicine* 68 (1997): 895–99.

Savours, Ann M. "Obituary: Mary C. Lobban." *Polar Record* 21 (1983): 403.

Schaffer, Simon. "The Asiatic Enlightenment of British Astronomy." In *The Brokered World: Go-Betweens and Global Intelligence, 1770–1820*, edited by S. Schaffer, L. Roberts, K. Raj, and J. Delbourgo, 49–104. Sagamore Beach, MA: Watson Publishing, 2009.

——. "The Information Order of Isaac Newton's *Principia Mathematica*." Hans Rausing Lecture, Uppsala University, Sweden, 2008.

Schaffner, Anna Katharina. *Exhaustion: A History.* New York: Columbia University Press, 2016.

Scheffler, Robin Wolfe. "The Fate of a Progressive Science: The Harvard Fatigue Laboratory, Athletes, the Science of Work and the Politics of Reform." *Endeavour* 35 (June 2011): 48–54.

Schiebinger, Londa. *Plants and Empire: Colonial Bioprospecting in the Atlantic World.* Cambridge, MA: Harvard University Press, 2004.

——. "Why Mammals Are Called Mammals: Gender Politics in Eighteenth-Century Natural History." *American Historical Review* 98 (1993): 382–411.

Schneider, W. "Blood Transfusion in Peace and War, 1900–1918." *Social History of Medicine* 10 (1997): 105–26.

Scott, Shirley V. "Ingenious and Innocuous? Article IV of the Antarctic Treaty as Imperialism." *Polar Journal* 1 (2011): 51–62.

Seag, Morgan. "Equal Opportunities on Ice: Examining Gender and Institutional Change at the British Antarctic Survey, 1975–1996." Master's thesis, University of Cambridge, 2015.

———. "Women Need Not Apply: Gendered Institutional Change in Antarctica and Outer Space." *Polar Journal* 7 (2017): 319–35.

Sera-Shriar, Efram. "Arctic Observers: Richard King, Monogenism and the Historicisation of Inuit through Travel Narratives." *Studies in History and Philosophy of Science Part C* 51 (2015): 23–31.

Shapin, Steven. "The Invisible Technician." *American Scientist* 77 (1989): 554–63.

Shelesnyak, M. C. "The History of the Arctic Research Laboratory, Point Barrow, Alaska." *Arctic* 1 (1948): 97–106.

Shephard, Roy J. "Work Physiology and Activity Patterns of Circumpolar Eskimos and Ainu: A Synthesis of IBP Data." *Human Biology* 46 (1974): 263–94.

Simons, E., and O. Oelz. "Mont Blanc with Oxygen: The First Rotters." *High Altitude Medicine & Biology* 2 (2001): 545–49.

Siple, Paul A. "Clothing and Climate." In *The Physiology of Heat Regulation and the Science of Clothing; Prepared at the Request of the Division of Medical Sciences, National Research Council*, edited by L. H. Newburgh, 433–41. Philadelphia: W. B. Saunders, 1949.

———. "General Principles Governing Selection of Clothing for Cold Climates." *Proceedings of the American Philosophical Society* 89 (1945): 200–234.

Siple, Paul A., and Charles F. Passel. "Measurements of Dry Atmospheric Cooling in Subfreezing Temperatures." *Proceedings of the American Philosophical Society* 89 (1945): 177–99.

Siri, William E., and Ann Lage. *William E. Siri: Reflections on the Sierra Club, the Environment and Mountaineering, 1950s–1970s*. Sierra Club History Series. Berkeley: Regional Oral History Office, Bancroft Library, University of California, 1979.

Smith, S. D. "Coffee, Microscopy, and the *Lancet's* Analytical Sanitary Commission." *Social History of Medicine* 14 (2001): 171–97.

Smith Hughes, Sally. "Interview Transcript: Will Siri." Bancroft Library, University of California Berkeley (1980). http://digitalassets.lib.berkeley.edu/rohoia/ucb/text/nuclear medicine00lawrrich.pdf.

Somervell, T. Howard. *After Everest: The Experiences of a Mountaineer and Medical Missionary.* 2nd ed. London: Hodder & Stoughton, 1939.

———. "Note on the Composition of Alveolar Air at Extreme Heights." *Journal of Physiology* 60 (1925): 282–85.

Sorrenson, Richard. "The Ship as a Scientific Instrument in the Eighteenth Century." *Osiris* 11 (1996): 221–36.

Stefansson, Vilhjalmur. *The Fat of the Land* (enlarged edition of *Not by Bread Alone*). New York: Macmillan, 1960.

———. *The Friendly Arctic: The Story of Five Years in Polar Regions.* New York: Macmillan, 1922.

———. *My Life with the Eskimo.* New York: Macmillan, 1912.

Stefansson, Vilhjalmur, and US War Department. *Arctic Manual.* Washington, DC: Government Printing Office, 1940.

Stepan, Nancy. *The Idea of Race in Science: Great Britain 1800–1960.* Houndmills, Hampshire: Macmillan, 1987.

Strasser, Bruno J. "Collecting, Comparing, and Computing Sequences: The Making of Margaret O. Dayhoff's Atlas of Protein Sequence and Structure, 1954–1965." *Journal of the History of Biology* 43 (2009): 623–60.

———. "The Experimenter's Museum: GenBank, Natural History, and the Moral Economies of Biomedicine." *Isis* 102 (2011): 60–96.

———. "Laboratories, Museums, and the Comparative Perspective: Alan A. Boyden's Serological Taxonomy, 1925–1962." *Historical Studies in the Natural Sciences* 40 (2010): 533–64.

Stroud, M. A. "Nutrition and Energy Balance on the 'Footsteps of Scott' Expedition 1984–86." *Human Nutrition: Applied Nutrition* 41 (1987): 426–33.

Sutton, John R. "A Lifetime of Going Higher: Charles Snead Houston." *Journal of Wilderness Medicine* 3 (1992): 225–31.

Sutton, John R., J. T. Reeves, P. D. Wagner, B. M. Groves, A. Cymerman, M. K. Malconian, P. B. Rock, P. M. Young, S. D. Walter, and C. S. Houston. "Operation Everest II: Oxygen Transport during Exercise at Extreme Simulated Altitude." *Journal of Applied Physiology* 64 (1988): 1309–21.

Taylor, Anthony J. W., and Iain A. McCormick. "Human Experimentation during the International Biomedical Expedition to the Antarctic (IBEA)." *Journal of Human Stress* 11 (1985): 161–64.

Thomson, M. L. "The Cause of Changes in Sweating Rate after Ultraviolet Radiation." *Journal of Physiology* 112 (1951): 31–42.

Tilman, H. W. *The Ascent of Nanda Devi.* Cambridge: Cambridge University Press, 1937.

Tracy, Sarah. "The Physiology of Extremes: Ancel Keys and the International High Altitude Expedition of 1935." *Bulletin of the History of Medicine* 86 (2012): 627–60.

Tuckey, Harriet Pugh. *Everest—The First Ascent: The Untold Story of Griffith Pugh, the Man Who Made It Possible.* London: Rider, 2013.

Turchetti, Simone, Simon Naylor, Katrina Dean, and Martin Siegert. "On Thick Ice: Scientific Internationalism and Antarctic Affairs, 1957–1980." *History and Technology* 24 (2008): 351–76.

United States Air Force. *German Aviation Medicine, World War II.* Vol. 2. Washington, DC: Government Printing Office, 1950.

Unsworth, Walt. *Everest: The Mountaineering History.* 3rd ed. London: Bâton Wicks, 2000.

Vetter, Jeremy, ed. *Knowing Global Environments: New Historical Perspectives on the Field Sciences.* New Brunswick, NJ: Rutgers University Press, 2011.

Viault, F. "Sur l'augmentation considérable de nombre des globules rouges dans le sang chez les habitants des hautes plataux de l'Amérique du Sud." *Comptes rendues de l'Académie des Sciences* 111 (1890): 917–18.

Ward, Michael. "The Descent from Makalu, 1961, and Some Medical Aspects of High Altitude Climbing." *Alpine Journal* 68 (1963): 11–19.

———. "The Height of Mount Everest." *Alpine Journal* 100 (1995): 30–33.

———. "Himalayan Scientific Expedition 1960–61 (A Himalayan Winter, Rakpa Peak, Ama Dablam, Makalu)." *Alpine Journal* 66 (1961): 343–64.

Warren, C. B. "The Medical and Physiological Aspects of the Mount Everest Expeditions." *Geographical Journal* 90 (1937): 126–43.

Weindling, Paul. *John W. Thompson: Psychiatrist in the Shadow of the Holocaust.* Rochester, NY: University of Rochester Press, 2010.

West, John B. "American Medical Research Expedition to Everest, 1981." *Himalayan Journal* 39 (1981–82): 19–25.

———. "Barcroft's Bold Assertion: All Dwellers at High Altitudes Are Persons of Impaired Physical and Mental Powers." *Journal of Physiology* 594 (2016): 1127–34.

———. "Barometric Pressures on Mt. Everest." *Journal of Applied Physiology* 86 (1999): 1062–66.

———. "Diffusing Capacity of the Lung for Carbon Monoxide at High Altitude." *Journal of Applied Physiology* 17 (1962): 421–26.

———. *Everest: The Testing Place.* New York: McGraw-Hill, 1985.

———. "George I. Finch and His Pioneering Use of Oxygen for Climbing at Extreme Altitudes." *Journal of Applied Physiology* 94 (2003): 1702–13.

———. *High Life: A History of High-Altitude Physiology and Medicine.* New York: Published for the American Physiological Society by Oxford University Press, 1998.

———. "Letter from Chowri Kang." *High Altitude Medicine & Biology* 2 (2001): 311–13.

———. "Times Past: Failure on Everest: The Oxygen Equipment of the Spring 1952 Swiss Expedition." *High Altitude Medicine & Biology* 4 (2003): 39–43.

West, John B., P. H. Hackett, K. H. Maret, J. S. Milledge, R. M. Peters Jr., C. J. Pizzo, and R. M. Winslow. "Pulmonary Gas Exchange on the Summit of Mount Everest." *Journal of Applied Physiology* 55 (1983): 678–87.

Wilkins, D. C. "Heat Acclimatization in the Antarctic." *Journal of Physiology* 214 (1971): 15–16.

Williams, E. S. "Sleep and Wakefulness at High Altitudes." *British Medical Journal* (January 24, 1959): 197–98.

Wilson, J. G. "The Himalayan Schoolhouse Expeditions." *Alpine Journal* 70 (1965): 226–39.

Wilson, Ken, and Mike Pearson. "Post-Mortem of an International Expedition." *Himalayan Journal* 31 (1971): 33–83.

Wilson, Ove. "Human Adaptation to Life in Antarctica." In *Biogeography and Ecology in Antarctica*, edited by J. van Mieghem and P. van Oye, 690–752. The Hague: W. Junk, 1965.

———. "Physiological Changes in Blood in the Antarctic." *British Medical Journal* 2 (December 26, 1953): 1425–28.

Windsor, Jeremy, Roger C. McMorrow, and George W. Rodway. "Oxygen on Everest: The Development of Modern Open-Circuit Systems for Mountaineers." *Aviation, Space, and Environmental Medicine* 79 (2008): 799–804.

Windsor, Jeremy, and George W. Rodway. "Heights and Haematology: The Story of Haemoglobin at Altitude." *Postgraduate Medical Journal* 83 (2007): 148–51.

Winslow, John B. "High-Altitude Polycythemia." In *High Altitude and Man*, edited by John B. West and Sukhamay Lahiri, 163–73. Bethesda, MD: American Physiological Society, 1984.

Worboys, Michael. "The Emergence of Tropical Medicine: A Study in the Establishment of a Scientific Specialty." In *Perspectives on the Emergence of Scientific Disciplines*, edited by Gerard Lemaine, 76–98. The Hague: De Gruyter Mouton, 1976.

———. "Tropical Medicine." In *Companion Encyclopaedia of the History of Medicine*, edited by Roy Porter and W. F. Bynum, 512–36. London: Taylor & Francis, 1993.

Wright, John. "British Polar Expeditions 1919–39." *Polar Record* 26 (1990): 77–84.

———. "The Polar Eskimos." *Polar Record* 3 (1939): 115–29.

Wrynn, Alison M. "'A Debt Was Paid Off in Tears': Science, IOC Politics and the Debate about High Altitude in the 1968 Mexico City Olympics." *International Journal of the History of Sport* 23 (2006): 1152–72.

Wulsin, Frederick R. "Adaptations to Climate among Non-European Peoples." In *The Physiology of Heat Regulation and the Science of Clothing; Prepared at the Request of the Division of Medical Sciences, National Research Council*, edited by L. H. Newburgh, 1–50. Philadelphia: W. B. Saunders, 1949.

Zak, Annie. "Obama Signs Measure to Get Rid of the Word 'Eskimo' in Federal Laws." *Alaska Dispatch News*, May 24, 2016. https://www.adn.com/alaska-news/2016/05/23/obama-signs-measure-to-get-rid-of-the-word-eskimo-in-federal-laws/.

Zink, R. A., W. Schaffert, K. Messmer, and W. Brendel. "Hemodilution: Practical Experiences in High Altitude Expeditions." In *High Altitude Physiology and Medicine*, edited by W. Brendel and R. A. Zink, 291–97. New York: Springer-Verlag, 1982.

Zuntz, N., A. Loewy, F. Muller, and W. Caspari. *Höhenklima und Bergwanderungen in ihrer Wirkung auf den Menschen*. Berlin: Deutsches Verlagshaus, 1906.

INDEX